W9-BXZ-786

Experiments in Circuit Analysis to Accompany Introductory Circuit Analysis

6th Edition

Robert L. Boylestad
Gabriel Kousourou

Merrill, an imprint of
Macmillan Publishing Company
New York

Collier Macmillan Canada, Inc.
Toronto

Maxwell Macmillan International Publishing Group
New York   Oxford   Singapore   Sydney

NOTICE TO THE READER

Cover Photo: Marcel Eskenazy/Photo Edit

Macmillan Publishing Company
866 Third Avenue, New York, NY 10022

Collier Macmillan Canada, Inc.

This book was set in Times Roman and Lubalin Graph.

Administrative Editor: David Garza
Developmental Editor: Carol Thomas
Production Editor: Rex Davidson
Art Coordinator: Pete Robison
Cover Designer: Cathy Watterson and Brian Deep
Text Designer: Cynthia Brunk

International Standard Book Number: 0–675–21182–4
Printed in the United States of America
2 3 4 5 6 7 8 9—94 93 92 91

To our wives and children

Merrill's International Series in Electrical and Electronics Technology

Preface

On the basis of the laboratory experience at our own institution, a number of laboratory experiments from the 5th edition were modified in this edition. It became apparent that the first few laboratory experiments of the dc section should be expanded to ensure a correct, clear understanding of a few fundamental meter operations. The first four experiments are now encompassed in six experiments providing additional time to investigate a variety of configurations and ensure a correct application of the DMM and/or VOM for resistance, voltage, and current measurements.

A number of experiments, such as Superposition and Thevenin's Theorem, have been rewritten primarily for increased clarity to ensure that the data requested is properly entered in the various tables.

In the ac section a number of frequency-response experiments were added to clearly demonstrate the effect of frequency on the characteristics of an ac network. In addition, the component values of the parallel resonant circuit were modified somewhat to provide an improved set of data.

In both the dc and ac sections two computer laboratory experiments were added to permit a confirmation of results obtained in earlier experiments. PSPICE was emphasized, but the opportunity to use any language remains.

The number of experiments in both the dc and ac sections has been increased to permit a choice of procedures for the typical dc-ac sequence. As with the 5th edition, a solutions manual containing results obtained during the testing of the experiments will be available to instructors.

Any reaction to the experiments of this edition and the introduction of computer laboratory exercises is requested by the authors. We trust you will find the content of this edition a measurable improvement over past editions.

We wish to thank Stephen Helba, David Garza, Carol Thomas, and Rex Davidson at Merrill Publishing Co. for their support and guidance in coordinating the production of this manual.

Robert L. Boylestad
Gabriel Kousourou

Contents

dc EXPERIMENTS

ac EXPERIMENTS

Math Review

dc

The analysis of dc circuits requires that a number of fundamental mathematical operations be performed on an ongoing basis. Most of the operations are typically covered in a secondary-level education; a few others will be introduced in the early sessions of the lecture portion of this course.

The material to follow is a brief review of the important mathematical procedures along with a few examples and exercises. Derivations and detailed explanations are not included but are left as topics for the course syllabus or as research exercises for the student.

It is strongly suggested that the exercises be performed with and without a calculator. Although the modern-day calculator is a marvelous development, the student should understand how to perform the operations in the long-hand fashion. Any student of a technical area should also develop a high level of expertise with the use of the calculator through frequent application with a variety of operations.

FRACTIONS

Addition and Subtraction

Addition and subtraction of fractions require that each term have a common denominator. A direct method of determining the common denominator is simply to multiply the numerator and denominator of each fraction by the denominator of the other fractions. The method of "selecting the least common denominator" will be left for a class lecture or research exercise.

EXAMPLES

$$\frac{1}{2} + \frac{2}{3} = \left(\frac{3}{3}\right)\left(\frac{1}{2}\right) + \left(\frac{2}{2}\right)\left(\frac{2}{3}\right) = \frac{3}{6} + \frac{4}{6} = \frac{7}{6}$$

$$\frac{3}{4} - \frac{2}{5} = \left(\frac{5}{5}\right)\left(\frac{3}{4}\right) - \left(\frac{4}{4}\right)\left(\frac{2}{5}\right) = \frac{15}{20} - \frac{8}{20} = \frac{7}{20}$$

$$\frac{5}{6} + \frac{1}{2} - \frac{3}{4} = \left(\frac{2}{2}\right)\left(\frac{4}{4}\right)\left(\frac{5}{6}\right) + \left(\frac{6}{6}\right)\left(\frac{4}{4}\right)\left(\frac{1}{2}\right) - \left(\frac{6}{6}\right)\left(\frac{2}{2}\right)\left(\frac{3}{4}\right)$$

$$= \frac{40}{48} + \frac{24}{48} - \frac{36}{48} = \frac{28}{48} = \frac{7}{12}$$

(12 being the least common denominator for the three fractions)

Multiplication

The multiplication of two fractions is straightforward, with the product of the numerator and denominator formed separately as shown below.

EXAMPLES

$$\left(\frac{1}{3}\right)\left(\frac{4}{5}\right) = \frac{(1)(4)}{(3)(5)} = \frac{4}{15}$$

$$\left(\frac{3}{7}\right)\left(\frac{1}{2}\right)\left(-\frac{4}{5}\right) = \frac{(3)(1)(-4)}{(7)(2)(5)} = -\frac{12}{70} = -\frac{6}{35}$$

Division

Division requires that the denominator be inverted and then multiplied by the numerator.

EXAMPLE

$$\frac{\dfrac{2}{3}}{\dfrac{4}{5}} = \frac{2}{3} \times \frac{5}{4} = \frac{10}{12} = \frac{5}{6}$$

Simplest Form

A fraction can be reduced to simplest form by finding the largest common divisor of both the numerator and denominator. As an illustration,

$$\frac{12}{60} = \frac{12/12}{60/12} = \frac{1}{5}$$

If the largest divisor is not obvious, start with a smaller common divisor, such as shown here:

$$\frac{12/6}{60/6} = \frac{2}{10} = \frac{2/2}{10/2} = \frac{1}{5}$$

Or, if you prefer, use repeated divisions of smaller common divisors, such as 2 or 3:

$$\frac{12/2}{60/2} = \frac{6}{30} = \frac{6/2}{30/2} = \frac{3}{15} = \frac{3/3}{15/3} = \frac{1}{5}$$

EXAMPLES

$$\frac{180}{300} = \frac{180/10}{300/10} = \frac{18}{30} = \frac{18/6}{30/6} = \frac{3}{5}$$

$$\frac{72}{256} = \frac{72/8}{256/8} = \frac{9}{32}$$

or

$$\frac{72}{256} = \frac{72/2}{256/2} = \frac{36}{128} = \frac{36/2}{128/2} = \frac{18}{64} = \frac{18/2}{64/2} = \frac{9}{32}$$

Mixed Numbers

Numbers in the mixed form (whole and fractional part) can be converted to the fractional form by multiplying the whole number by the denominator of the fractional part and adding the numerator.

EXAMPLES

$$3\frac{4}{5} = \frac{(3 \times 5) + 4}{5} = \frac{19}{5}$$

$$60\frac{1}{3} = \frac{(60 \times 3) + 1}{3} = \frac{181}{3}$$

The reverse process is simply a division operation with the remainder left in fractional form.

Conversion to Decimal Form

Fractions can be converted to decimal form by simply performing the indicated division.

EXAMPLES

$$\frac{1}{4} = 4\overline{)1.00}^{\,0.25} = 0.25$$

$$\frac{3}{7} = 7\overline{)3.0000}^{\,0.4285\ldots} = 0.4285\ldots$$

$$4\frac{1}{5} = 4 + 5\overline{)1.0}^{\,0.2} = 4.2$$

PROBLEMS

Perform each of the following operations by hand and compare to the calculator solution.

1. $\dfrac{3}{5} + \dfrac{1}{3} =$ _____ (by hand) _____ (calculator)

2. $\dfrac{5}{9} - \dfrac{1}{4} =$ _____ (by hand) _____ (calculator)

3. $\dfrac{9}{10} - \dfrac{1}{2} + \dfrac{1}{4} =$ _____ (by hand) _____ (calculator)

4. $\left(\dfrac{5}{6}\right)\left(\dfrac{9}{13}\right) =$ _____ (by hand) _____ (calculator)

5. $\left(-\dfrac{1}{7}\right)\left(+\dfrac{8}{15}\right)\left(-\dfrac{3}{5}\right) =$ _____ (by hand) _____ (calculator)

6. $\dfrac{5/6}{2/3} =$ _____ (by hand) _____ (calculator)

7. $\dfrac{9/10}{-6/7} =$ _____ (by hand) _____ (calculator)

8. Reduce to simplest form:

   a. $\dfrac{72}{84} =$ _____ (by hand) _____ (calculator)

   b. $\dfrac{144}{384} =$ _____ (by hand) _____ (calculator)

9. Convert to the mixed form:

   a. $\dfrac{16}{7}$ = _____

   b. $\dfrac{320}{9}$ = _____

10. Convert to decimal form:

   a. $\dfrac{3}{8}$ = _____ (by hand) _____ (calculator)

   b. $6\dfrac{5}{6}$ = _____ (by hand) _____ (calculator)

SCIENTIFIC NOTATION

The need to work with very small and large numbers requires that the use of powers of 10 be appreciated and clearly understood.

    The direction of the shift of the decimal point and the number of places moved will determine the resulting power of 10. For instance:

$$10{,}000. = 1 \times 10^{+4} = 10^{+4}$$

$$5000. = 5 \times 10^{+3}$$

$$0.000001 = 10^{-6}$$

$$0.00456 = 4.56 \times 10^{-3}$$

Addition and Subtraction

Addition or subtraction of numbers using scientific notation requires that the power of 10 of each term be the same.

EXAMPLES

$$45{,}000 + 3000 + 500 = 45 \times 10^3 + 3 \times 10^3 + 0.5 \times 10^3 = 48.5 \times 10^3$$

$$0.02 - 0.003 + 0.0004 = 200 \times 10^{-4} - 30 \times 10^{-4} + 4 \times 10^{-4}$$
$$= (200 - 30 + 4) \times 10^{-4} = 174 \times 10^{-4}$$

Multiplication

Multiplication using powers of 10 employs the following equation, where a and b can be any positive or negative number:

$$\boxed{10^a \times 10^b = 10^{a+b}} \qquad \text{(M.1)}$$

EXAMPLES

$$(100)(5000) = (10^2)(5 \times 10^3) = 5 \times 10^2 \times 10^3 = 5 \times 10^{2+3} = 5 \times 10^5$$

$$(200)(0.0004) = (2 \times 10^2)(4 \times 10^{-4}) = (2)(4)(10^2)(10^{-4}) = 8 \times 10^{2-4} = 8 \times 10^{-2}$$

Note in the above examples that the operations with powers of 10 can be separated from the integer values.

Division

Division employs the following equation, where a and b can again be any positive or negative number:

$$\boxed{\frac{10^a}{10^b} = 10^{a-b}} \qquad \text{(M.2)}$$

EXAMPLES

$$\frac{320,000}{4000} = \frac{32 \times 10^4}{4 \times 10^3} = \frac{32}{4} \times 10^{4-3} = 8 \times 10^1 = 80$$

$$\frac{1600}{0.0008} = \frac{16 \times 10^2}{8 \times 10^{-4}} = 2 \times 10^{2-(-4)} = 2 \times 10^{2+4} = 2 \times 10^6$$

Note in the last example the importance of carrying the proper sign through Eq. M.2.

Powers

Powers of powers of 10 can be determined using the following equation, where a and b can be any positive or negative number:

$$\boxed{(10^a)^b = 10^{ab}} \qquad \text{(M.3)}$$

EXAMPLES

$$(1000)^4 = (10^3)^4 = 10^{12}$$

$$(0.002)^3 = (2 \times 10^{-3})^3 = (2)^3 \times (10^{-3})^3 = 8 \times 10^{-9}$$

PROBLEMS

Using powers of 10, perform the following operations by hand, and then compare to the calculator solution.

11. $5,800,000 + 450,000 + 2000 = $ _____ (by hand) _____ (calculator)

12. $0.04 + 0.008 - 0.3 = $ _____ (by hand) _____ (calculator)

13. $2400 + 0.05 \times 10^3 - 40,000 \times 10^{-3} = $ _____ (by hand) _____ (calculator)

14. $(68,000)(40,000) = $ _____ (by hand) _____ (calculator)

15. $(0.0009)(0.006) = $ _____ (by hand) _____ (calculator)

16. $(-5 \times 10^{-8})(8 \times 10^5)(-0.02 \times 10^4) = $ _____ (by hand) _____ (calculator)

17. $\dfrac{0.00081}{0.009} = $ _____ (by hand) _____ (calculator)

18. $\dfrac{5000}{6 \times 10^6} = $ _____ (by hand) _____ (calculator)

19. $\dfrac{-8 \times 10^{-4}}{0.002} = $ _____ (by hand) _____ (calculator)

20. $(4 \times 10^5)^3 = $ _____ (by hand) _____ (calculator)

21. $(0.0003)^4 = $ _____ (by hand) _____ (calculator)

22. $[(4000)(0.0003)]^3 = $ _____ (by hand) _____ (calculator)

PREFIXES

The frequent use of some powers of 10 has resulted in their being assigned abbreviations that can be applied as prefixes to a numerical value in order to quickly identify its relative magnitude.

The following is a list of the most frequently used prefixes in electrical and electronic technology:

$$10^6 = \text{mega (M)}$$
$$10^3 = \text{kilo (k)}$$
$$10^{-3} = \text{milli (m)}$$
$$10^{-6} = \text{micro } (\mu)$$
$$10^{-9} = \text{nano (n)}$$
$$10^{-12} = \text{pico (p)}$$

EXAMPLES

$$6,000,000\,\Omega = 6 \text{ megohms} = 6\,M\Omega$$
$$0.04\,A = 40 \times 10^{-3}\,A = 40\,mA$$
$$0.00005\,V = 50 \times 10^{-6}\,V = 50\,\mu V$$

When converting from one form to another, be aware of the relative magnitude of the quantity before you start to provide a check once the conversion is complete. Starting out with a relatively small quantity and ending up with a result of large relative magnitude clearly indicates an error in the conversion process. In many cases the best method may be to return to the decimal form before establishing the new form for the quantity.

EXAMPLES

$$0.008\,M\Omega = 0.008 \times 10^6\,\Omega = 8 \times 10^3\,\Omega = 8\,k\Omega$$
$$5600\,mA = 5600 \times 10^{-3}\,A = 5.6\,A$$
$$20,000\,mV = 20,000 \times 10^{-3}\,V = 20\,V = 0.02\,kV$$

PROBLEMS

Apply the most appropriate prefix to each of the following quantities.

23. $0.00006\,A =$ _____

24. $504,000\,\Omega =$ _____

25. $32,000\,V =$ _____

26. $0.000000009\,A =$ _____

Perform the following conversions.

27. $35\,mA =$ _____ A

28. $0.005\,kV =$ _____ V = _____ mV

29. $8,000,000\,\Omega =$ _____ MΩ = _____ kΩ

30. $4000\,pF =$ _____ nF = _____ μF

OTHER FUNCTIONS

The student will encounter a variety of other mathematical functions in the study of electrical and electronic systems. A few will now be examined to insure their correct evaluation when the need arises.

Square and Cubic Root of a Number

The calculator will always be used to determine the square and cubic roots of a number. In each case, the power of 10 must be divisible by 2 or 3, respectively, or the resulting power of 10 will not be a whole number.

EXAMPLES

$$\sqrt{200} = (200)^{1/2} = (2 \times 10^2)^{1/2} = (2)^{1/2} \times (10^2)^{1/2} \cong 1.414 \times 10^1 = 14.14$$
$$\sqrt{0.004} = (0.004)^{1/2} = (40 \times 10^{-4})^{1/2} = (40)^{1/2} \times (10^{-4})^{1/2}$$
$$= 6.325 \times 10^{-2} = 0.06325$$
$$\sqrt[3]{5000} = (5000)^{1/3} = (5 \times 10^3)^{1/3} = (5)^{1/3} \times 10^1 \cong 1.71 \times 10^1 = 17.1$$

Some calculators provide the $\sqrt[3]{}$ function while others require that you use the $\sqrt[x]{y}$ function. For some calculators, the y^x function is all that is available, requiring that the y value be entered first, followed by the y^x function and then the power ($x = 1/3 = 0.3333$).

PROBLEMS

31. $\sqrt{6.4} =$ _____
32. $\sqrt[3]{3000} =$ _____
33. $\sqrt{0.00005} =$ _____

EXPONENTIAL FUNCTIONS

The exponential functions e^x and e^{-x} will appear frequently in the analysis of $R\text{-}C$ and $R\text{-}L$ networks with switched dc inputs (or square-wave inputs). On most calculators, only e^x is provided, requiring that the user insert the negative sign when necessary.

Keep in mind that e^x is equivalent to $(2.71828\ldots)^x$ or a number to a power x. For any positive values of x greater than 1, the result is a magnitude greater than $2.71828\ldots$ In other words, for increasing values of x, the magnitude of e^x will increase rapidly. For $x = 0$, $e^0 = 1$, and for increasing negative values of x, e^{-x} will become increasingly smaller.

When inserting the negative sign for e^{-x} functions, be sure to use the proper key to enter the negative sign. It is not always the key used for addition.

EXAMPLES

$$e^{+2} \cong 7.3890$$
$$e^{+10} \cong 22026.47$$
$$e^{-2} \cong 0.13534$$
$$e^{-10} \cong 0.0000454$$

PROBLEMS

34. $e^{+4.2} =$ _____
35. $e^{+0.5} =$ _____
36. $e^{-0.02} =$ _____

ALGEBRAIC MANIPULATIONS

The following is a brief review of some basic algebraic manipulations that must be performed in the analysis of dc circuits. It is assumed that a supporting math course will expand on the coverage provided here.

EXAMPLES

(a) Given $v = \dfrac{d}{t}$, solve for d and t.

Both sides of the equation must first be multiplied by t:

$$(t)(v) = \left(\frac{d}{\cancel{t}}\right)(\cancel{t})$$

resulting in $d = vt$.

Dividing both sides of $d = vt$ by v yields

$$\frac{d}{v} = \frac{\cancel{v}t}{\cancel{v}}$$

and

$$t = \frac{d}{v}$$

In other words, the proper choice of multiplying factors for both sides of the equation will result in an equation for the desired quantity.

(b) Given $R_1 + 4 = 3R_1$, solve for R_1.

In this case, $-R_1$ is added to both sides, resulting in

$$(R_1 + 4) - R_1 = 3R_1 - R_1$$

and

$$4 = 2R_1$$

with

$$R_1 = \frac{4}{2} = 2\Omega$$

(c) Given $\frac{1}{R_1} = \frac{1}{3R_1} + \frac{1}{6}$, solve for R_1.

Multiplying both sides of the equation by R_1 results in

$$\frac{\cancel{R_1}}{\cancel{R_1}} = \frac{\cancel{R_1}}{3\cancel{R_1}} + \frac{R_1}{6}$$

or

$$1 = \frac{1}{3} + \frac{R_1}{6}$$

Subtracting 1/3 from both sides gives

$$1 - \frac{1}{3} = \frac{1}{3} - \frac{1}{3} + \frac{R_1}{6}$$

and

$$\frac{2}{3} = \frac{R_1}{6}$$

with

$$R_1 = \frac{(6)(2)}{3} = \frac{12}{3} = 4\Omega$$

PROBLEMS

37. Given $I = \dfrac{E}{R}$, solve for R and E.

$R =$ _____ , $E =$ _____

38. Given $30I = 5I + 5$, solve for I.

$I =$ _____

39. Given $\dfrac{1}{R_T} = \dfrac{1}{R_1} + \dfrac{1}{4R_1}$, solve for R_T in terms of R_1.

$R_T =$ _____

40. Given $F = \dfrac{kQ_1Q_2}{r^2}$, solve for Q_1 and r.

$Q_1 =$ _____ , $r =$ _____

Resistors and the Color Code

OBJECT

To become familiar with the use of the ohmmeter and the resistor color code.

EQUIPMENT REQUIRED

Resistors

1 — 1-MΩ, 1-W carbon resistor

1 — 1-MΩ, 2-W carbon resistor

1 — 91-Ω, 220-Ω, 2.2-kΩ, 10-kΩ, 1-MΩ, 1/2-W carbon resistors

Instruments

1 — VOM (Volt-Ohm-Milliammeter)

1 — DMM (Digital Multimeter)

EQUIPMENT ISSUED

TABLE 1.1

Item	Manufacturer and Model No.	Laboratory Serial No.
VOM		
DMM		

DESCRIPTION OF EQUIPMENT

Both the VOM and the DMM will be used in this experiment to measure the resistance of the provided resistors. The VOM employs an analog scale to read resistance, voltage, and current, while the DMM has a digital display. An analog scale is a continuous scale, requiring that the user be able to interpret the location of the pointer using the scale divisions provided. For resistance measurements, the analog scale is also nonlinear, resulting in smaller distances between increasing values of resistance. Note on the VOM scale the relatively large distance between 1 Ω and 10 Ω and the smaller distance between 100 Ω and 1000 Ω. The digital display provides a numerical value with the accuracy determined by the chosen scale. For many years the analog meter was the instrument employed throughout the industry. In recent years the digital meter has grown in popularity, making it important that the graduate of any technical program be adept at using both types of meters.

RÉSUMÉ OF THEORY

In this experiment the resistance of a series of 1/2-W carbon resistors will first be determined from the color code and then compared with the measured value using both the VOM and the DMM. The two meters are applied to ensure familiarity with reading both an analog and a digital scale.

The procedure for determining the resistance of a color-coded resistor is described in Appendix I with a listing of the numerical value associated with each color. The first two bands (those closest to the end of the resistor) determine the first two digits of the resistor value, while the third band determines the power of the power-of-10 multiplier (actually the number of zeros to follow the first two digits). The fourth band is the percent tolerance for the chosen resistor (see Fig. 1.1).

Red Red Br Silver

2   2   1   10%

FIG. 1.1

$220 \pm 10\% = 220 \pm 22 = 198\ \Omega$ to $242\ \Omega$

For increasing wattage ratings, the size of the carbon resistor will increase to provide the "body" required to dissipate the resulting heating effects.

Resistance is *never* measured by an ohmmeter in a *live* network, due to the possibility of damaging the meter with excessively high currents and obtaining readings that have no meaning. It is usually best to remove the resistor from the circuit before measuring its resistance to ensure that the measured value does not include the effect of other resistors in the system. However, if this maneuver is undesirable or impossible, make sure that one end of the resistor is not connected to any other element.

The use of the VOM and DMM will be described early in the laboratory session. For the VOM, always reset the zero-adjust whenever you change scales. In addition, always choose the dial setting (R × 1, R × 10, and so on) that will place the pointer in the region of the scale that will give the best reading. Finally, do not forget to multiply the reading by the proper multiplying factor. For the DMM, remember that any scale marked "kΩ" will be reading in kilohms, and any "MΩ" scale in megohms. For instance, a 91-Ω resistor may read 90.7 Ω on a 200-Ω scale, 0.091 on a 2-kΩ scale, and 0.000 on a 2-MΩ scale. There is no zero-adjust on a DMM meter, but make sure that $R = 0 \; \Omega$ when the leads are touching or an adjustment internal to the meter may have to be made. Any resistance above the maximum for a chosen scale will result in an O.L. indication.

One last comment is in order. Remember that there is no polarity to resistance measurements. Either lead of the meter can be placed on either end of the resistor — the resistance will be the same. Polarities will become important, however, when we measure voltages and currents in the experiments to follow.

PROCEDURE

The purpose of this first experiment is to acquaint you with the equipment, so *do not rush*. Learn how to read the meter scales accurately, and take your data carefully. If you are uncertain about anything, do not hesitate to ask your instructor.

Part 1 Body Size

(a) In the space below, draw the physical sizes of 1/2-W, 1-W, and 2-W, 1-MΩ carbon resistors (color bands = brown, black, green). Note in particular that the resistance of each is the same but that the size increases with wattage rating.

(b) How much larger (approximately) is the 1-W resistor than the 1/2-W resistor? Is the ratio the same for the 2-W resistor as compared with the 1-W resistor?

Part 2 Color Code

(a) Complete Table 1.2 for each of the provided 1/2-W resistors. Examine the resistors in any order, but remember that resistor 1 in Table 1.2 is resistor 1 in the other tables to follow. An example is provided to define the procedure.

TABLE 1.2

Resistor	Color Bands – Color				Color Bands – Numerical Value				Nominal Resistance
	1	2	3	4	1	2	3	4	
Sample	Red	Red	Black	Gold	2	2	0	5%	22 Ω
1									
2									
3									
4									
5									

The percent tolerance is used to determine the range of resistance levels within which the manufacturer guarantees the resistor will fall. It is determined by first taking the percent tolerance and multiplying by the nominal resistance level. For the example in Table 1.2, the resulting resistance level

$$(5\%)(22 \ \Omega) = (0.05)(22 \ \Omega) = 1.1 \ \Omega$$

is added to and subtracted from the nominal value to determine the range as follows:

$$\text{Maximum value} = 22 \ \Omega + 1.1 \ \Omega = 23.1 \ \Omega$$
$$\text{Minimum value} = 22 \ \Omega - 1.1 \ \Omega = 20.9 \ \Omega$$

as shown in Table 1.3.

(b) Complete Table 1.3 for each resistor in Table 1.2.

TABLE 1.3

Resistor	Maximum Resistance	Minimum Resistance
Sample	23.1 Ω	20.9 Ω
1		
2		
3		
4		
5		

The resistance levels will now be read using the VOM and DMM and inserted in Table 1.4. The percent difference will be determined using the following equation:

$$\% \text{ Difference} = \frac{|\text{Measured} - \text{Nominal}|}{\text{Nominal}} \times 100\%$$

For the sample and the DMM reading:

$$\% \text{ Difference} = \frac{|22.9 \ \Omega - 22 \ \Omega|}{22 \ \Omega} \times 100\%$$

$$= \frac{0.9}{22} \times 100\% = \underline{\underline{4.09\%}}$$

TABLE 1.4

Resistor	Measured Value – VOM	Falls Within Specified Tolerance (Yes/No)	Measured Value – DMM	Falls Within Specified Tolerance (Yes/No)	% Difference (Using DMM Reading)
Sample	23 Ω	Yes	22.9 Ω	Yes	4.09
1					
2					
3					
4					
5					

Part 3　Body Resistance

Guess the resistance of your body between your hands and record the value in Table 1.5. Measure the resistance with the DMM (or VOM) by firmly holding one lead in each hand (wetting your fingers will improve the reading), and record in the same table. If 10 mA are "lethal," what voltage ($V = IR$) would be required to produce the current through your body? Again, record in Table 1.5.

TABLE 1.5

Guessed body resistance	
Measured body resistance	
Lethal voltage	

Calculations:

Part 4　Meter Resistance

(a) Voltmeters　Ideally, the internal resistance of the DMM and VOM should be infinite (like an open circuit) when voltages in a network are being measured to ensure that the meter does not alter the normal behavior of the network.

For the VOM, there is an ohm/volt (Ω/V) rating written on the bottom of the scale that permits determination of the internal resistance of each scale of the meter when used as a voltmeter. For the 260 Simpson VOM, the ohm/volt rating is 20,000 Ω/V. The internal resistance of each setting can then be calculated by multiplying the maximum voltage reading of a scale by the ohm/volt rating. For instance, the 10-V scale will have (10 V)(20,000 Ω/V) = 200 kΩ, while the 250-V scale will have (250 V)(20,000 Ω/V) = 5 MΩ. No matter what voltage level is measured on each scale, the internal resistance will remain the same.

Using the DMM as an ohmmeter, measure the resistance of each voltage scale of the VOM and record in Table 1.6. Insert the maximum voltage levels of your meter if different from those appearing in the table.

TABLE 1.6

VOM	Scales* of Your VOM	Ω/V Rating	Calculated Resistance	Measured Resistance	% Difference $= \dfrac{\lvert \text{Meas.} - \text{Calc.}\rvert}{\text{Calc.}} \times 100\%$
2.5 V					
10 V					
50 V					
250 V					
500 V					

*If different from those in the first column.

Most DMMs have the same internal resistance for each scale. In other words, the internal resistance of a typical meter will be 10 MΩ whether the 2-V or the 200-V scale is used.

Use the VOM to measure the internal resistance of each scale of the DMM and complete Table 1.7. Make sure the DMM is on.

TABLE 1.7

DMM	Your DMM* (Insert Scale Settings)	Specified Internal Resistance	Measured Resistance
200 mV			
2 V			
20 V			
200 V			
1500 V			

*If different from those in the first column.

In general, which meter would appear to disturb the network less?

(b) Ammeters The ammeter is also an instrument that when inserted in a network should not adversely affect the normal current levels. However, since it is placed in series with the branch in which the current is being measured, it should ideally have a resistance of zero ohms (or as small as possible).

Using the DMM as an ohmmeter, measure the resistance of each current scale of the VOM and record in Table 1.8. Insert the maximum current levels of your meter if different from those appearing in the table.

TABLE 1.8

VOM	Your VOM* (Insert Scale Settings)	Measured Resistance
1 mA		
10 mA		
100 mA		
500 mA		

*If different from those in the first column.

Using the VOM as an ohmmeter, measure the resistance of each current scale of the DMM and record in Table 1.9. As above, insert the maximum current levels of your meter if different from those appearing in the table.

TABLE 1.9

DMM	Your DMM* (Insert Scale Settings)	Measured Resistance
2 mA		
20 mA		
200 mA		
2 A		

*If different from those in the first column.

Which meter would appear to disturb the network the least?

PROBLEMS

1. What are the ohmic values and tolerances of the following carbon resistors?

Color Bands — Color				Numerical Value	Tolerance
1	**2**	**3**	**4**		
Brown	Black	Blue	Gold		
Yellow	Violet	Orange	Gold		
Brown	Gray	Gold	None		
Red	Yellow	Silver	Gold		
Green	Brown	Green	Silver		
Green	Blue	Black	None		

2. For the VOM, which region of the scale normally provides the best readings? Why?

Ohm's Law

OBJECT

To become familiar with the use of the laboratory dc supply and with measuring voltages and currents

EQUIPMENT REQUIRED

Resistors

1 – 1000-Ω, 10,000-Ω, 1/2-W

Instruments

1 – DMM (or VOM)
1 – dc Power supply

EQUIPMENT ISSUED

TABLE 2.1

Item	Manufacturer and Model No.	Laboratory Serial No.
DMM		
VOM		
Power supply		

TABLE 2.2

Resistors	
Nominal (Color-Coded) Value	Measured Value
1,000 Ω	
10,000 Ω	

RÉSUMÉ OF THEORY

In any active circuit there must be a source of power. In the laboratory, it is convenient to use a source that requires a minimum of maintenance and, more important, whose output voltage can be varied easily. Power supplies are rated as to maximum voltage and current output. For example, a supply rated 0–40 V at 500 mA will provide a maximum voltage of 40 V and a maximum current of 500 mA at any voltage.

Most dc power supplies have three terminals, labeled as shown in Fig. 2.1. The

FIG. 2.1

three terminals permit the establishment of a positive or negative voltage, which can be grounded or ungrounded. The variable voltage is available only between terminals *A* and *B*. Both *A* and *B* must be part of any connection scheme. If only terminals *A* and *B* are employed as shown in Fig. 2.2, the supply is considered "floating" and not connected to the common ground of the network. For common ground and safety reasons, the supply is normally grounded as shown in Fig. 2.3 for a positive voltage, and as in Fig. 2.4 for a negative voltage.

FIG. 2.2

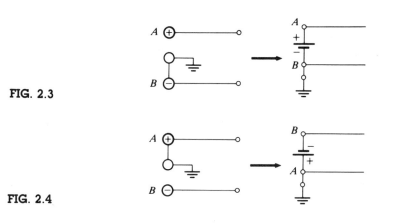

FIG. 2.3

FIG. 2.4

When measuring voltage levels, make sure the voltmeter is connected in parallel (across) the element being measured as shown in Fig. 2.5. In addition, recognize that if the leads are connected as shown in the figure, the reading will be up-scale and positive. If the meter were hooked up in the reverse manner, a negative (down-scale, below-zero) reading would result. The voltmeter is therefore an excellent instrument for not only measuring the voltage level but also determining the polarity. Since the meter is always placed in parallel with the element, there is no need to disturb the network when the measurement is made.

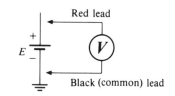

FIG. 2.5

Ammeters are always connected in series with the branch in which the current is being measured as shown in Fig. 2.6, normally requiring that the branch be opened and the meter inserted. Ammeters also have polarity markings to indicate the manner in which they should be connected to obtain an up-scale reading. Since the current I of Fig. 2.6 would establish a voltage drop across the ammeter as illustrated, the reading of the ammeter will be up-scale and positive. If the meter were hooked up in the reverse manner, the reading would be negative or down-scale. In other words, simply reversing the leads will change a below-zero indication to an up-scale reading.

FIG. 2.6

Until you become familiar with the use of the ammeter, draw in the ammeter in the network with the polarities determined by the current direction. It is then easier to ensure that the meter is connected properly to the surrounding elements. This process will be demonstrated in more detail in a later experiment.

For both the voltmeter and the ammeter, always start with the higher ranges and work down to the operating level to avoid damaging the instrument. When the VOM and DMM are returned to the stockroom, be sure the VOM is on the highest voltage scale and the DMM is in the off position.

The voltage across and the current through a resistor can be used to determine its resistance using Ohm's law in the following form:

$$R = \frac{V}{I}$$ (2.1)

The magnitude of R will be determined by the units of measure for V and I.

PROCEDURE

The purpose of this laboratory exercise is to acquaint you with the equipment, so *do not rush*. If you are a member of a squad, don't let one individual make all the measurements. You must become comfortable with the instruments if you expect to perform your future job function in a professional manner. Read the instruments carefully. The more accurate a reading, the more accurate the results obtained. One final word of caution. For obvious reasons, *do not make network changes with the power on!* If you have any questions about the procedure, be sure to contact your instructor.

Part 1

Hook up the DMM (or VOM) to the power supply as shown in Fig. 2.7.

FIG. 2.7

Using the DMM, adjust the power supply to the voltages listed in Table 2.3. Read the voltage V_{AB} at each level using the DMM and VOM.

TABLE 2.3

V_{AB}	DMM	VOM
2 V		
4 V		
6 V		
8 V		
10 V		

Comment on the readings obtained with each meter. Which was easier to use? Which appeared to be more accurate? Answer the questions in some detail.

Part 2 Ohm's Law

(a) Construct the circuit of Fig. 2.8

FIG. 2.8

Insert the measured value of R from Table 2.2 and use it for all calculations. Note that the voltmeter is placed in parallel with the resistor with the positive (red) lead connected to the point of higher potential as established by the indicated current direction.

Adjust the power supply until $E = V = 2$ V using the voltmeter. Then turn off the power and construct the circuit of Fig. 2.9 to measure the current I. Note the manner in which the meter is connected to ensure that the meter polarities match those established by the current direction. Record the value of I in Table 2.4.

FIG. 2.9

TABLE 2.4

	$R = 1\ k\Omega$	
$E = V$	I	$R = \dfrac{V}{I}$
2 V		
4 V		
6 V		
8 V		
10 V		

(b) Repeat the above process for a range of values for E from 2 V to 10 V in 2-V steps and insert the measured value of I for each step in Table 2.4.

(c) Calculate the resistance using the supply voltage and measured current and Ohm's Law and complete the second column. Show all calculations below:

(d) How do the calculated values of R in the second column compare with the nameplate value of the resistor (1 kΩ)?

(e) How do the calculated values of R in the second column compare with the measured value (by an ohmmeter) appearing in Fig. 2.8?

(f) Plot the data from Table 2.4 on Graph 2.1 using the scales provided.

(g) The slope of the curve is related to the resistance by

$$\boxed{\text{Slope} = m = \frac{1}{R}} \qquad (2.2)$$

In other words, the less the resistance the steeper the slope and vice versa. The resistance R can be determined from the graph using the equation $R = V/I$ at any point in the graph.

Determine the resistance of the resistor plotted on Graph 2.1 at $I = 2.6$ mA and at $V = 8.2$ V.

$I = 2.6$ mA:

GRAPH 2.1

$V = 8.2$ V:

(h) The resistance can also be determined from the equation

$$R = \frac{\Delta V}{\Delta I}$$ (2.3)

where ΔI is the corresponding change in current due to a change in voltage ΔV or vice versa.

Determine the resistance of the resistor plotted on Graph 2.1 using a $\Delta V = 6\ V - 2\ V = 4\ V$

Using the resistance level just calculated, find the slope m of the curve using Eq. (2.2).

(i) Complete Table 2.5 for $R = 10\ k\Omega$ using the same procedure described for parts 2(a) and 2(b)

TABLE 2.5

E	I	$R = \dfrac{V}{I}$
	$R = 10\ k\Omega$	
2 V		
4 V		
6 V		
8 V		
10 V		

(j) Calculate the resistance using the supply voltage and measured current and Ohm's Law and complete the second column of Table 2.5. Show all calculations below:

(k) How do the calculated values of R in the second column of Table 2.5 compare with the nameplate value of the resistor (10 kΩ)?

(l) How do the calculated values of R in the second column of Table 2.5 compare with the measured value (by an ohmmeter) of the 10-kΩ resistor?

(m) Plot the data from Table 2.5 on Graph 2.1 using the scales provided.

(n) How would you compare the slopes of the curves?

(o) Do the results of part 2(n) substantiate Eq. (2.2)?

(p) Determine the resistance at $V = 5.6$ V and compare with the measured value.

$R =$ _____

(q) Determine the resistance using a $\Delta V = 6$ V $- 2$ V $= 4$ V and Eq. (2.3) and compare the results with the measured value.

$R =$ _____

PROBLEMS

1. Plot the linear curve for a 100-Ω resistor on Graph 2.1 and comment on the slope compared with those for the other resistors.

2. Sketch the proper hookup for two parallel grounded supplies providing 10 V at twice the current rating of either supply.

3. Why does the equation $R = V/I$ provide the same result as the equation $R = \Delta V/\Delta I$ for any level of V or ΔV for the fixed carbon composition resistors?

Series Resistance

OBJECT

To determine the total resistance of series dc circuits.

EQUIPMENT REQUIRED

Resistors

1 — 220-Ω, 330-Ω, 1-kΩ, 1-MΩ

3 — 100-Ω

Instruments

1 — DMM (or VOM)

1 — dc Power supply

EQUIPMENT ISSUED

TABLE 3.1

Item	Manufacturer and Model No.	Laboratory Serial No.
DMM (or VOM)		
Power supply		

TABLE 3.2

Resistors	
Nominal Value	**Measured Value**
100 Ω	
100 Ω	
100 Ω	
220 Ω	
330 Ω	
1 kΩ	
1 MΩ	

RÉSUMÉ OF THEORY

The total resistance R_T of a series circuit is the sum of the individual resistances. For Fig. 3.1,

$$R_T = R_1 + R_2 + R_3 \tag{3.1}$$

FIG. 3.1

For equal series resistors, the total resistance is equal to the resistance of one resistor times the number in series. That is,

$$R_T = NR \tag{3.2}$$

For an unknown system, such as that appearing in Fig. 3.2, the total resistance can be determined by the following form of Ohm's law:

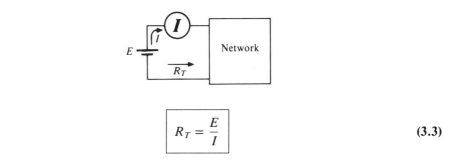

FIG. 3.2

$$R_T = \frac{E}{I} \tag{3.3}$$

That is, for an applied voltage E the current is measured and Eq. (3.3) applied to determine the resistance of the network.

For resistors of difference magnitudes in series, the total resistance is determined primarily by the largest resistors. In some instances it is very close to the value of one resistor if that resistor is much larger than the others.

PROCEDURE

Part 1   Two Series Resistors

(a) Construct the circuit of Fig. 3.3.

FIG. 3.3

In the spaces provided, insert the measured value of each resistor as determined using an ohmmeter.

(b) Calculate the total resistance using the measured resistor values.

$R_T =$ _____

(c) Measure the total resistance using the ohmmeter section of your DMM or VOM.

$R_T =$ _____

How do the calculated and measured values compare? Calculate the percent difference from

$$\% \text{ Difference} = \frac{|\text{Measured} - \text{Calculated}|}{\text{Calculated}} \times 100\% \qquad (3.4)$$

% Difference = _____

(d) Apply 10 V to the circuit and insert an ammeter as shown in Fig. 3.4. Set the supply voltage with the voltmeter section of your meter and then turn off the supply. Insert the meter in the ammeter mode as shown and turn on the supply. Record the ammeter reading below.

$I = $ _____

FIG. 3.4

(e) Using the supply voltage and ammeter reading of part 1(d), calculate the total resistance using Eq. (3.3).

$R_T = $ _____

How do the results of this part and part 1(c) compare?

Part 2   Three Series Resistors

(a) Construct the circuit of Fig. 3.5. Insert the measured values of the resistors in the spaces provided.

FIG. 3.5

(b) Calculate the total resistance using the measured resistor values.

$R_T =$ _____

(c) Measure the total resistance using the ohmmeter section of your DMM or VOM.

$R_T =$ _____

How do the calculated and measured values compare? Calculate the percent difference using Eq. (3.4).

% Difference = _____

(d) Apply 10 V and an ammeter in the same manner described in part 1(d) and record the source current below.

$I =$ _____

(e) Using the supply voltage and ammeter reading of part 2(d), calculate the total resistance using Eq. (3.3).

$R_T =$ _____

How do the results of this part and part 2(c) compare?

Part 3   Equal Series Resistors

(a) Construct the circuit of Fig. 3.6. Insert the measured values of the resistors in the spaces provided.

FIG. 3.6

(b) Using the nominal resistor values (all 100 Ω), calculate the total resistance using Eq. (3.2).

$R_T = $ _____

(c) Measure the total resistance using the ohmmeter section of your DMM or VOM.

$R_T = $ _____

How do the calculated and measured values compare? Calculate the percent difference using Eq. (3.4).

% Difference = _____

(d) Calculate the total resistance using the measured values of each resistor.

$R_T = $ _____

How does this calculated value compare with the measured value of part 3(c)? What is the percent difference?

% Difference = _____

Part 4   A Composite

(a) Construct the circuit of Fig. 3.7. Insert the measured resistor values in the spaces provided.

FIG. 3.7

(b) Calculate the total resistance using the measured values.

$R_T =$ _____

(c) Measure the total resistance and find the percent difference between the calculated and measured values.

$R_T =$ _____

% Difference = _____

(d) Apply 10 V and measure the source current I. Then calculate the total resistance and compare with the measured value of part 4(c).

$I =$ _____

$R_T =$ _____

% Difference = _____

Part 5 Different Levels of Resistance

(a) Construct the circuit of Fig. 3.8. Insert the measured resistor values in the spaces provided.

FIG. 3.8

(b) Using the measured values, calculate the total resistance of the circuit.

$R_T = $ _____

(c) Using an ohmmeter, measure the total resistance.

$R_T = $ _____

Calculate the percent difference between the measured and calculated values.

% Difference = _____

(d) Calculate the total resistance using measured values if the resistance R_1 is ignored.

$R_T = $ _____

What is the percent difference between this R_T and the measured value of part 5(c)?

% Difference = _____

(e) What is the total resistance using the measured value if both the resistances R_1 and R_2 are ignored.

$R_T =$ _____

What is the percent difference between this R_T and the measured value of part 5(c)?

% Difference = _____

(f) Noting the percent differences of parts 5(d) and 5(e), what approximations do you think can be made when calculating the total resistance of Fig. 3.8?

Series dc Circuits

OBJECT

To investigate the characteristics of a series dc circuit

EQUIPMENT REQUIRED

Resistors

1 — 330-Ω, 470-Ω

2 — 220-Ω

Instruments

1 — DMM (or VOM)

1 — dc Power supply

EQUIPMENT ISSUED

TABLE 4.1

Item	Manufacturer and Model No.	Laboratory Serial No.
DMM (or VOM)		
Power supply		

TABLE 4.2

Resistors	
Nominal Value	Measured Value
220 Ω	
220 Ω	
330 Ω	
470 Ω	

RÉSUMÉ OF THEORY

In a series circuit, the current is the same through all of the circuit elements, as shown by the current I in Fig. 4.1.

FIG. 4.1

By Ohm's law, the current I is $I = E/R_T$, where E is the impressed voltage and R_T the total resistance. Applying Kirchhoff's voltage law around the closed loop of Fig. 4.1, we find

$$\boxed{E = V_1 + V_2 + V_3}$$

(4.1)

with $V_1 = IR_1$, $V_2 = IR_2$, and $V_3 = IR_3$.

The voltage divider rule states that the voltage across an element or across a series combination of elements in a series circuit is equal to the resistance of the element or series combination of elements divided by the total resistance of the series circuit and multiplied by the total impressed voltage. For the elements of Fig. 4.1,

$$V_1 = \frac{R_1 E}{R_T} = \frac{R_1 E}{R_1 + R_2 + R_3} \ , \quad V_2 = \frac{R_2 E}{R_T} \ , \quad V_3 = \frac{R_3 E}{R_T}$$

PROCEDURE

Part 1 Series Resistance Measurements

(a) Construct the circuit shown in Fig. 4.2.

Record the measured value of each resistor in the place indicated. Use the DMM (or VOM) to set $E = 20$ V.

R_1 measured = _____
R_2 measured = _____
R_3 measured = _____

FIG. 4.2

Measure the voltage across the supply and across each resistor with the DMM (or VOM), making sure to note the polarity of the voltage across each element. Record the voltages below.

$E =$ _____ , $V_1 =$ _____ , $V_2 =$ _____ ,

$V_3 =$ _____

From your readings of the voltages, is Kirchhoff's voltage law satisfied around path $ABCD$? That is, does $E = V_1 + V_2 + V_3$?

(b) Now calculate the current through each resistor and record (use measured resistor values).

$I_1 = \dfrac{V_1}{R_1} =$ _____ , $I_2 = \dfrac{V_2}{R_2} =$ _____ ,

$I_3 = \dfrac{V_3}{R_3} =$ _____

How do the values of I_1, I_2, and I_3 compare?

What can you deduce about the characteristics of a series circuit from the current calculations above?

(c) The current *I* can be measured by inserting the multimeter (in the milliammeter mode) in series with the source as shown in Fig. 4.3.

FIG. 4.3

Measure the current at positions *A*, *B*, *C*, and *D* of Fig. 4.3 and record below.

I(at A) = _____ , I(at B) = _____ , I(at C) =

_____ , I(at D) = _____

Do your results verify the calculations of part 1(b)?

(d) Compute the total resistance of the circuit using the following equation:

$$R_T = \frac{E}{I}$$

R_T = _____

(e) Using the measured values of the resistors from Table 4.2, calculate the total resistance R_T, where $R_T = R_1 + R_2 + R_3$. Record the value.

R_T = _____

(f) Shut down and disconnect the power supply. Then measure the input resistance (R_T) across points A-D using the DMM (or VOM). Record that value.

R_T (measured) = _____

Compare this value with those in parts 1(d) and 1(e).

(g) Using the voltage divider rule, calculate the value of the voltage across resistor R_2, and then compare this value with the measured value from part 1(a).

V_2 (voltage divider rule) = _____ ,

V_2 [from part 1(a)] = _____

Part 2   Voltage Divider Rule

Construct the circuit of Fig. 4.4.

FIG. 4.4

(a) Calculate V_1 using the voltage divider rule and measured resistance values.

V_1 = _____

(b) Measure V_1 and compare with the calculated value of part 2(a).

$V_1 = $ _____

(c) Measure V_2 and V_3 and comment on whether equal resistors in series have equal voltage drops across them.

$V_2 = $ _____ , $V_3 = $ _____

(d) Calculate the voltage V using the voltage divider rule.

$V = $ _____

(e) Measure the voltage V and compare with the value of part 2(d).

$V = $ _____

(f) Using measured resistor values, calculate the current I.

$I = $ _____

(g) Measure the current I and compare with the result of part 2(f).

$I = $ _____

PROBLEMS

1. Given the voltage V_1 in Fig. 4.5, determine the resistor R_1 using the voltage divider rule.

FIG. 4.5

$$R_1 = \underline{\hspace{2cm}}$$

2. Determine V_1 for the configuration of Fig. 4.6.

FIG. 4.6

$$V_1 = \underline{\hspace{2cm}}$$

Parallel Resistance

OBJECT

To determine the total resistance of parallel dc circuits.

EQUIPMENT REQUIRED

Resistors

$1-100\text{-}\Omega$, $1\text{-}k\Omega$, $1.2\text{-}k\Omega$, $2.2\text{-}k\Omega$, $1\text{-}M\Omega$
$3-3.3\text{-}k\Omega$

Instruments

$1-$ DMM (or VOM)
$1-$ dc Power supply

EQUIPMENT ISSUED

TABLE 5.1

Item	Manufacturer and Model No.	Laboratory Serial No.
DMM (or VOM)		
Power supply		

TABLE 5.2

Resistors	
Nominal Value	**Measured Value**
100 Ω	
1 kΩ	
1.2 kΩ	
2.2 kΩ	
3.3 kΩ	
3.3 kΩ	
3.3 kΩ	
1 MΩ	

RÉSUMÉ OF THEORY

In a parallel circuit (Fig. 5.1), the total or equivalent resistance (R_T) is given by

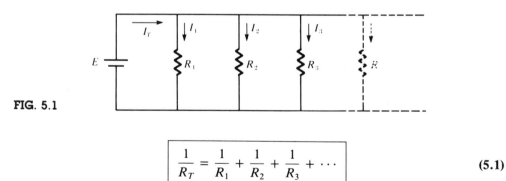

FIG. 5.1

$$\frac{1}{R_T} = \frac{1}{R_1} + \frac{1}{R_2} + \frac{1}{R_3} + \cdots \qquad (5.1)$$

If there are only two resistors in parallel, it is more convenient to use

$$R_T = \frac{R_1 R_2}{R_1 + R_2} \qquad (5.2)$$

For equal parallel resistors, the total resistance is equal to the resistance of one resistor divided by the number of resistors in parallel. That is,

$$R_T = \frac{R}{N}$$ **(5.3)**

For an unknown system the same technique applied for series dc circuits can be applied to parallel dc circuits. In Fig. 5.2 the total resistance can be determined from

$$R_T = \frac{E}{I}$$ **(5.4)**

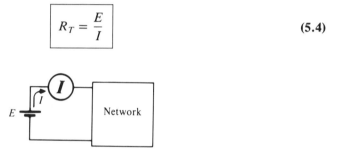

FIG. 5.2

For parallel resistors, the total resistance is always less than the value of the smallest parallel resistor. In some instances the total resistance is very close to the value of one resistor if that resistance is much less than the others.

PROCEDURE

Part 1   Two Parallel Resistors

(a) Construct the circuit of Fig. 5.3.

FIG. 5.3

Insert the measured values of the resistors in the spaces provided.

(b) Using the measured resistor values, calculate the total resistance using Eq. (5.1). Show all work.

$R_T =$ _____

(c) Using the ohmmeter section of your DMM or VOM, measure the total resistance of the network of Fig. 5.3.

$R_T =$ _____

(d) How do the results of parts 1(b) and 1(c) compare? Calculate their percent difference from

$$\% \text{ Difference} = \frac{|\text{Measured} - \text{Calculated}|}{\text{Calculated}} \times 100\% \qquad (5.5)$$

% Difference = _____

(e) Calculate the total resistance using measured values and Eq. (5.2). How do the results compare with part 1(b)?

$R_T =$ _____

(f) Apply 10 V to the circuit and insert the milliammeter section of the DMM (or VOM) as shown in Fig. 5.4. Set the supply voltage with the voltmeter section of your meter and then insert the meter in the milliammeter mode as shown. Record the milliammeter reading below.

$I =$ _____

FIG. 5.4

(g) Using the supply voltage and the milliammeter reading of part 1(f), calculate the total resistance using Eq. (5.4).

$R_T =$ _____

How do the results of this part and part 1(c) compare?

(h) How close is the total resistance to the resistance of the smaller resistor of Fig. 5.3? Is it less than R_1, and is it close to the magnitude of R_1?

Part 2   Three Parallel Resistors

(a) Construct the network of Fig. 5.5.

FIG. 5.5

Insert the measured values of the resistors in the spaces provided.

(b)   Calculate the total resistance using the measured resistor values.

$R_T =$ _____

(c) Measure the total resistance and calculate the percent difference.

$R_T =$ _____

% Difference = _____

(d) Apply 10 V and insert the milliammeter section of the DMM (or VOM) as shown in Fig. 5.2, and record the source current below.

$I =$ _____

(e) Using the supply voltage and the ammeter reading of part 2(d), calculate the total resistance using Eq. (5.4).

$R_T =$ _____

How do the results of this part and part 2(c) compare?

(f) How close is the total resistance to the resistance of the smallest parallel resistor? Is it less than R_1, and is it close to the magnitude of R_1?

Has the addition of the 3.3-kΩ resistor made the total resistance closer to R_1 than it was in part 1(h)? Can you generate a conclusion based on this result?

Part 3   Equal Parallel Resistors

(a) Construct the network of Fig. 5.6.

FIG. 5.6

$R_{1\,measured} =$ _____

$R_{2\,measured} =$ _____

$R_{3\,measured} =$ _____

Insert the measured resistor values in the spaces provided.

(b) Using the nominal resistor values (all 3.3 kΩ), calculate the total resistance using Eq. (5.3).

$R_T =$ _____

(c) Measure the total resistance using the ohmmeter section of your DMM or VOM.

$R_T =$ _____

How do the calculated and measured values compare? Calculate the percent difference using Eq. (5.5).

% Difference = _____

(d) Calculate the total resistance using the measured values of each resistor.

$R_T =$ _____

How does this calculated value compare with the measured value of part 3(c)? What is the percent difference?

% Difference = _____

Part 4   A Composite

(a) Construct the network of Fig. 5.7.

FIG. 5.7

$R_{1\,measured} =$ _____
$R_{2\,measured} =$ _____
$R_{3\,measured} =$ _____
$R_{4\,measured} =$ _____

Insert the measured resistor values in the spaces provided.

(b) Calculate the total resistance using the measured values.

$R_T =$ _____

(c) Measure the total resistance and find the percent difference between the calculated and measured values.

$R_T =$ _____

% Difference = _____

(d) Apply 10 V and measure the source current I. Then calculate the total resistance and compare with the measured value of part 4(c).

$I =$ _____

$R_T =$ _____

% Difference = _____

Part 5 Different Levels of Resistance

(a) Construct the network of Fig. 5.8.

FIG. 5.8

$R_{1 \, measured} =$ _____

$R_{2 \, measured} =$ _____

$R_{3 \, measured} =$ _____

Insert the measured resistor values in the spaces provided.

(b) Using the measured values, calculate the total resistance of the circuit.

$R_T =$ _____

(c) Using an ohmmeter, measure the total resistance.

$R_T =$ _____

Calculate the percent difference between the measured and calculated values.

% Difference = _____

(d) Calculate the total resistance using measured values if the resistance R_3 is ignored.

$R_T =$ _____

What is the percent difference between this R_T and the measured value of part 5(c)?

% Difference = _____

(e) What is the total resistance using the measured values if resistances R_2 and R_3 are ignored.

$R_T =$ _____

What is the percent difference between this R_T and the measured value of part 5(c)?

% Difference = _____

(f) It was noted in the Résumé of Theory that the total resistance of parallel resistors is always less than the value of the smallest resistor. Was this true in parts 5(b) and 5(c)?

Using the results of parts 5(d) and 5(e), what general conclusion can you come to regarding the total resistance of parallel resistors when one resistor is considerably smaller (10:1 or greater) than the other parallel resistors?

Parallel dc Circuits

OBJECT

To investigate the characteristics of parallel dc circuits

EQUIPMENT REQUIRED

Resistors

1 — 1.2-kΩ, 3.3-kΩ

2 — 2.2-kΩ

Instruments

1 — DMM (or VOM)

1 — dc Power supply

EQUIPMENT ISSUED

TABLE 6.1

Item	Manufacturer and Model No.	Laboratory Serial No.
DMM (or VOM)		
Power supply		

TABLE 6.2

Resistors	
Nominal Value	**Measured Value**
1.2 kΩ	
2.2 kΩ	
2.2 kΩ	
3.3 kΩ	

RÉSUMÉ OF THEORY

In a parallel circuit (Fig. 6.1), the voltage across parallel elements is the same. Therefore,

$$V_1 = V_2 = V_3 = E \tag{6.1}$$

FIG. 6.1

For the network of Fig. 6.1, the currents are related by the following expression:

$$I_S = I_1 + I_2 + I_3 + \cdots \tag{6.2}$$

and the current through each resistor is simply determined by Ohm's law:

$$I_1 = \frac{E}{R_1}, \; I_2 = \frac{E}{R_2}, \; I_3 = \frac{E}{R_3}, \text{ etc.} \tag{6.3}$$

The current divider rule (CDR) states that the current through one of two parallel branches is equal to the resistance of the *other* branch divided by the sum of the resistances of the two parallel branches and multiplied by the total current entering the two parallel branches. That is, for the network of Fig. 6.2,

FIG. 6.2

$$I_1 = \frac{R_2 I_S}{R_1 + R_2} \quad \text{and} \quad I_2 = \frac{R_1 I_S}{R_1 + R_2}$$ (6.4)

PROCEDURE

Part 1   Unequal Parallel Resistors

(a) Construct the network of Fig. 6.3. Insert the measured values of the resistors in the spaces provided.

$R_{1\,\text{measured}} = \text{_____}$

$R_{2\,\text{measured}} = \text{_____}$

FIG. 6.3

(b) Using the measured values, calculate the total resistance of the network.

$R_T(\text{calculated}) = \text{_____}$

(c) Measure R_T with the ohmmeter section and compare with the result of part 1(b).

R_T (measured) = _____

(d) How does the total resistance compare with that of the smaller of the two parallel resistors? Will this always be true?

(e) Apply 20 V to points 1-2 of Fig. 6.3 and verify that the voltage across each resistor is 20 V.

$V_{R_1} =$ _____ , $V_{R_2} =$ _____

Calculate the currents through R_1 and R_2 and determine the source current I_S.

$I_1 =$ _____ , $I_2 =$ _____ , $I_S =$ _____

(f) Measure I_S and compare with the result of part 1(e). Make sure the milliammeter is in series with the voltage source.

I_S (measured) = _____

Part 2 Equal Parallel Resistors

(a) Construct the network of Fig. 6.4. Insert the measured values of the resistors in the spaces provided.

FIG. 6.4

(b) Calculate the total resistance assuming exactly equal resistor values. Use the nominal color-coded value.

$R_T =$ _____

(c) Connect the ohmmeter to points 1-2 and record the reading below. Compare with the result of part 2(b).

$R_{\text{ohmmeter}} =$ _____

(d) Using the measured resistance levels of R_1 and R_2, determine the total resistance.

$R_T =$ _____

(e) Compare the value obtained in part 2(d) with that obtained in part 2(b). Would you recommend using measured (rather than color-coded nominal) values in the future?

(f) Apply 10 V to points 1-2 of Fig. 6.4 and verify that the voltage across R_1 and R_2 is 10 V.

$$V_{R_1} = \text{\underline{\hspace{2cm}}} \quad , \quad V_{R_2} = \text{\underline{\hspace{2cm}}}$$

(g) Calculate the currents I_1 and I_2 using Ohm's law, and determine I_S using Kirchhoff's current law.

$$I_1 = \text{\underline{\hspace{2cm}}} \quad , \quad I_2 = \text{\underline{\hspace{2cm}}} \quad , \quad I_S = \text{\underline{\hspace{2cm}}}$$

(h) Measure the currents I_1, I_2, and I_S using the milliammeter range of your meter and compare with the results of part 2(g).

I_1 (measured) = \underline{\hspace{2cm}} , I_2 (measured) = \underline{\hspace{2cm}} ,

I_S (measured) = \underline{\hspace{2cm}}

Part 3   Three Parallel Resistors

(a) Construct the network of Fig. 6.5. Insert the measured value of each resistor.

$R_{1\,measured} = \text{\underline{\hspace{1.5cm}}}$
$R_{2\,measured} = \text{\underline{\hspace{1.5cm}}}$
$R_{3\,measured} = \text{\underline{\hspace{1.5cm}}}$

FIG. 6.5

(b) Using the measured values, calculate the total resistance R_T.

R_T (calculated) = _____

(c) Determine R_T using the ohmmeter section of the multimeter and compare with the result of part 3(b).

R_T (measured) = _____

(d) Calculate the currents I_1, I_2, and I_3 if a source voltage of 10 V is applied. Calculate the source current I_S using Kirchhoff's current law.

$I_1 =$ _____ , $I_2 =$ _____ , $I_3 =$ _____

$I_S =$ _____

(e) Apply 10 V to the network of Fig. 6.5 and measure the currents I_3 and I_S. Compare with the results of part 3(d).

$I_3 =$ _____ , $I_S =$ _____

(f) Is the total resistance less than the magnitude of the smallest resistance? Comment accordingly.

Rheostats and Potentiometers

OBJECT

To study the characteristics and applications of a linear rheostat and potentiometer.

EQUIPMENT REQUIRED

Resistors

1 — 1-kΩ

1 — 0–1-kΩ linear carbon potentiometer

Instruments

1 — DMM (or VOM)

1 — dc Power supply

EQUIPMENT ISSUED

TABLE 7.1

Item	Manufacturer and Model No.	Laboratory Serial No.
DMM (or VOM)		
Power supply		

TABLE 7.2

Resistors	
Nominal Value	Measured Value
1 kΩ	

RÉSUMÉ OF THEORY

The rheostat is a two-terminal variable resistance device whose terminal characteristics can be linear (straight-line) or nonlinear. It is typically used in series with the load to control the voltage or current levels.

The potentiometer (often referred to as a *pot*) is a three-terminal device that is used primarily to control *potential* (voltage) levels. The potentiometer can reduce the voltage across a load to zero, while a two-terminal rheostat can reduce it only to a level determined by its maximum resistance and the resistance of the load.

A three-terminal potentiometer can be used as a two-terminal rheostat if desired. We will examine both possibilities in this experiment.

PROCEDURE

Part 1 Potentiometer Characteristics

(a) The graphic symbol for the potentiometer and the actual device appear in Fig. 7.1. As indicated by the schematic representation, the resistance between the outside terminals is fixed at the full value while the resistance between the movable contact and either outside terminal is a function of the position of the "wiper arm."

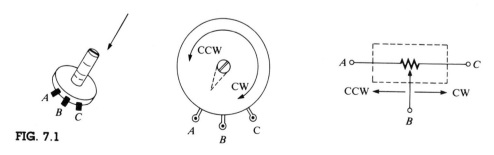

FIG. 7.1

(b) Turn the shaft to the full CW position and mark the position of the shaft on Fig. 7.2(b) as shown in Fig. 7.2(a).

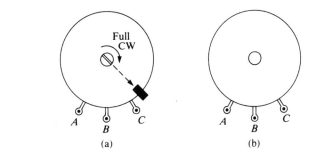

FIG. 7.2 (a) (b)

(c) Turn the shaft to the full CCW position and mark the position of the shaft on Fig. 7.2(b) as shown in Fig. 7.3.

FIG. 7.3

(d) Break down the region between full CW and full CCW of the potentiometer of Fig. 7.2(b) as demonstrated by Fig. 7.4. In other words, mark the locations of a 1/4, a 1/2, and a 3/4 turn from the full CCW position to the CW position. We will need these positions for the measurements to be performed.

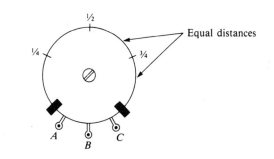

FIG. 7.4

(e) Connect the DMM (or VOM) to the terminals indicated in Table 7.3 and record the resistance value for each setting.

TABLE 7.3

Potentiometer Setting	Resistance To Be Measured		
	R_{AC}	R_{AB}	R_{BC}
Fully CCW (A)			
1/4 turn CW from A			
1/2 turn CW from A			
3/4 turn CW from A			
Fully CW (C)			

(f) Plot the resistances R_{AC}, R_{AB}, and R_{BC} versus position on Graph 7.1.

(g) Determine R_{AC} from $R_{AC} = R_{AB} + R_{BC}$ at the 1/4 turn position.

$R_{AC} = R_{AB} + R_{BC} = $ _____ + _____ = _____

Repeat the above for the 1/2 turn position.

$R_{AC} = $ _____

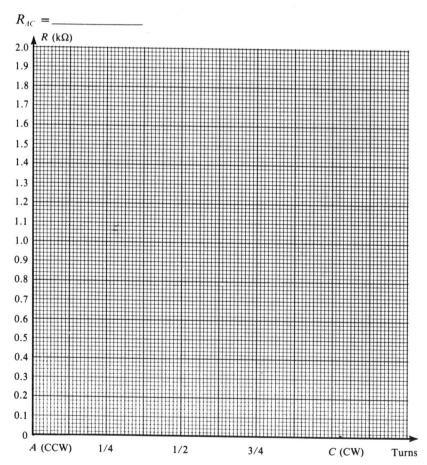

GRAPH 7.1

How do the values obtained above for R_{AC} compare with the measured values?

(h) Does R_{AC} (the resistance between the two outermost terminals) remain fairly constant throughout the measurement span? Why?

(i) Is the magnitude (ignore the sign) of the slope of R_{AB} the same as for R_{BC}? Should it be? Why?

(j) At 1/2 turn, does $R_{AB} = R_{BC}$? Should it be? Why?

Part 2   Potentiometer Control of Potential (Voltage) Levels

(a) Construct the network of Fig. 7.5.
Set the dc supply voltage with the DMM (or VOM).

FIG. 7.5

Using the VOM and DMM, read the voltages V_{CA}, V_{BA}, and V_{CB} at each step of the sequence indicated in Table 7.4. Be sure to read both voltages before changing the setting. Do not change the setting based on your readings. Leave the setting as determined by the judgments made on Fig. 7.2(b).

TABLE 7.4

Potentiometer Setting	V_{CA}	V_{BA}	V_{CB}
Fully CCW (A)			
1/4 turn CW from A			
1/2 turn CW from A			
3/4 turn CW from A			
Fully CW (C)			

    (b) Plot the voltages V_{CA}, V_{BA}, and V_{CB} versus position on Graph 7.2.

    (c) Which voltage remains independent of the position of the rotor? Why?

    (d) What are the minimum and maximum values of V_{BA} and V_{CB}?

Why do they have the above maximum and minimum values?

    (e) How would you compare the magnitudes of the slopes of the curves for V_{BA} and V_{CB}?

    (f) At the 5/8 turn position on Graph 7.2, does $V_{CA} = V_{CB} + V_{BA}$ using the graphical data?

$V_{CB} =$ _____

$V_{BA} =$ _____

$V_{CA} =$ _____

If it does, explain why.

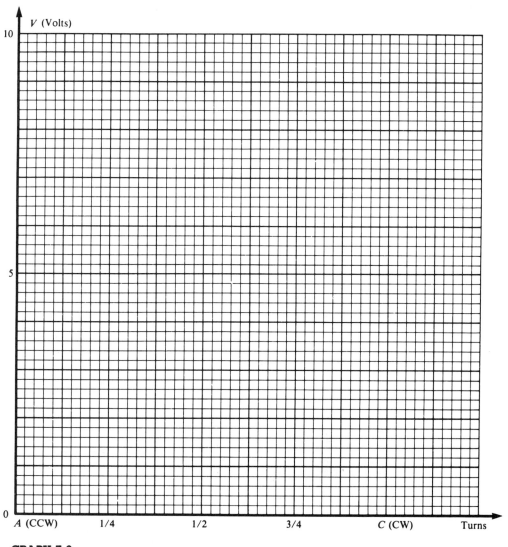

GRAPH 7.2

Part 3 Voltage Control

(a) Construct the circuit of Fig. 7.6.

Read the voltage V_{CB} at each step indicated in Table 7.5. Calculate the current I using the measured resistance value.

FIG. 7.6

TABLE 7.5

Potentiometer Setting	V_{CB}	$I = \dfrac{V_{CB}}{R_{\text{measured}}}$
Fully CCW (A)		
1/4 turn CW from A		
1/2 turn CW from A		
3/4 turn CW from A		
Fully CW (C)		

(b) Plot the voltage V_{CB} versus position on Graph 7.3.

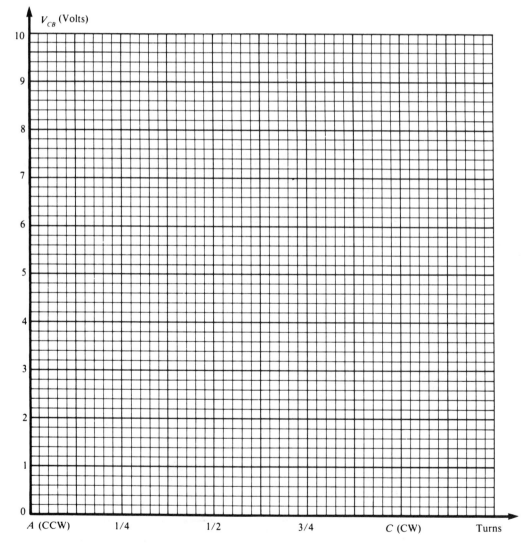

GRAPH 7.3

(c) Is the curve linear or nonlinear? Why?

(d) Plot the curve of the current I versus position on Graph 7.4.

(e) Does the potentiometer also offer a form of current control? Is this relationship linear or nonlinear?

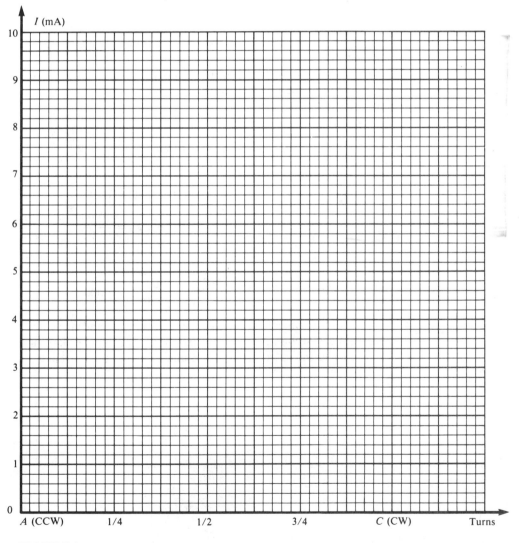

GRAPH 7.4

Part 4   Current Control with a Two-Point Rheostat

(a) Construct the network of Fig. 7.7.

Note that only two terminals of the potentiometer are being used. The total resistance of the circuit is now $R_{BC} + R$.

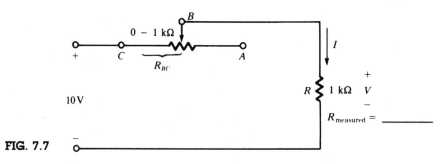

FIG. 7.7

(b) Set the rheostat to the minimum resistance level (fully CW) to establish 10 V across the 1-kΩ resistor. Then adjust the rheostat until the voltage across the 1-kΩ resistor is 9 V [using the DMM (or VOM)]. Calculate the resistance R_{BC} (using the measured values of the 1-kΩ resistor and the voltage divider rule) required to set this voltage level, and insert in Table 7.6. That is,

$$V = \frac{RE}{R + R_{BC}} = \frac{(1000)(10)}{1000 + R_{BC}}$$

or

$$(1000 + R_{BC})V = 10{,}000$$

$$1000 + R_{BC} = \frac{10{,}000}{V}$$

and

$$R_{BC} = \frac{10{,}000}{V} - 1000$$

Disconnect the power supply at point C and measure the resistance R_{BC} with the ohmmeter section of the meter and insert in Table 7.6. Repeat the above for the other voltage levels indicated in Table 7.6 and calculate the current I.

TABLE 7.6

V	R_{BC} (calculated)	R_{BC} (measured)	$I = \dfrac{V_R}{R_{\text{measured}}}$
10 V	0 Ω	0 Ω	10 mA
9 V			
8 V			
7 V			
6 V			
5 V			

(c) What is the minimum level the voltage V can be set to? Why is it limited to this level?

(d) Using the data obtained in Table 7.6, comment on the rheostat as a current control device.

(e) Are you satisfied with the way the calculated and measured values of R_{BC} compare? Comment accordingly.

Series-Parallel dc Circuits

OBJECT

To investigate the characteristics of series-parallel dc circuits.

EQUIPMENT REQUIRED

Resistors

1 — 91-Ω, 220-Ω, 330-Ω, 470-Ω

Instruments

1 — DMM (or VOM)

1 — dc Power supply

EQUIPMENT ISSUED

TABLE 8.1

Item	Manufacturer and Model No.	Laboratory Serial No.
DMM (or VOM)		
Power supply		

TABLE 8.2

Resistors	
Nominal Value	**Measured Value**
91 Ω	
220 Ω	
330 Ω	
470 Ω	

RÉSUMÉ OF THEORY

The analysis of series-parallel dc networks requires a firm understanding of the basics of both series and parallel networks. In the series-parallel configuration, you will have to isolate series and parallel configurations and make the necessary combinations for reduction as you work toward the desired unknown quantity.

As a rule, it is best to make a mental sketch of the path you plan to take toward the complete solution before introducing the numerical values. This may result in savings in both time and energy. Always work with the isolated series or parallel combinations in a branch before tying the branches together. For complex networks, a carefully redrawn set of reduced networks may be required to ensure that the unknowns are conserved and that every element has been properly included.

PROCEDURE

Part 1

(a) Construct the series-parallel network of Fig. 8.1, and insert the measured resistor values in the spaces provided.

FIG. 8.1

(b) Calculate the total resistance R_T using the measured resistance values.

R_T(calculated) = _____

(c) Use the ohmmeter section of the multimeter to measure R_T and compare with the result of part 1(b).

R_T(measured) = _____

(d) If 10 V were applied as shown in Fig. 8.2, *calculate* the currents I_S, I_1, I_2, and I_3 using the measured resistor values.

FIG. 8.2

(e) Apply 10 V and *measure* the currents I_1, I_2, and I_3 using the milliammeter section of your multimeter. Compare with the results of part 1(d). Be sure the meter is in series with the resistor through which the current is to be measured.

I_1(measured) = _____ , I_2(measured) = _____ ,

I_3(measured) = _____

(f) Using the results of part 1(d), *calculate* the voltages V_1, V_2, and V_3.

$V_1 =$ _____ , $V_2 =$ _____ , $V_3 =$ _____

(g) *Measure* the voltages V_1, V_2, and V_3 and compare with the results of part 1(f).

$V_1 =$ _____ , $V_2 =$ _____ , $V_3 =$ _____

(h) Referring to Fig. 8.2, does $E = V_1 + V_2$? Use the measured values of part 1(g) to check on this equality.

Part 2

(a) Construct the series-parallel network of Fig. 8.3.

(b) Calculate the total resistance R_T using the measured value of each resistor.

FIG. 8.3

R_T(calculated) = _____

(c) Use the ohmmeter section of your multimeter to measure the total resistance R_T. Compare with the result of part 2(b).

R_T(measured) = _____

(d) If 10 V were applied to the network as shown in Fig. 8.4, *calculate* the currents I_S, I_1, I_2, and I_3.

FIG. 8.4

$I_S =$ _____ , $I_1 =$ _____ , $I_2 =$ _____

$I_3 =$ _____

(e) Apply 10 V and measure the currents I_S, I_1, I_2, and I_3. Compare with the results of part 2(d).

I_S(measured) = _____ , I_1(measured) = _____ ,

I_2(measured) = _____ , I_3(measured) = _____

(f) Using the results of part 2(d), calculate the voltages V_1, V_2, and V_3.

$V_1 =$ _____ , $V_2 =$ _____ , $V_3 =$ _____

(g) Measure the voltages V_1, V_2, and V_3 and compare with the results of part 2(f).

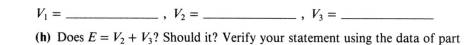

$V_1 =$ _____ , $V_2 =$ _____ , $V_3 =$ _____

(h) Does $E = V_2 + V_3$? Should it? Verify your statement using the data of part 2(g).

Part 3

(a) Construct the series-parallel network of Fig. 8.5 and insert the measured value of each resistor.

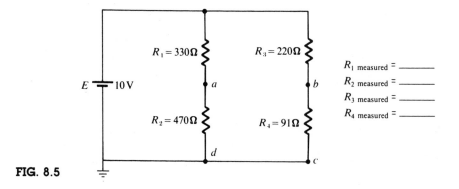

FIG. 8.5

(b) Using the measured resistor values, calculate the voltages V_{ad} and V_{bc}.

$V_{ad} =$ _____ , $V_{bc} =$ _____

(c) Measure the voltages V_{ad} and V_{bc} and compare with the results of part 3(b).

$V_{ad} =$ _____ , $V_{bc} =$ _____

(d) Using the results of part 3(b), calculate the voltage V_{ab}.

V_{ab}(calculated) = _____

(e) Measure the voltage V_{ab} and compare with the result of part 3(d).

V_{ab}(measured) = _____

Part 4

(a) Construct the network of Fig. 8.6. Insert the measured value of each resistor.

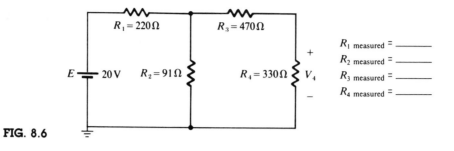

FIG. 8.6

(b) Calculate the voltage V_4 using the measured resistor values.

V_4(calculated) = _____

(c) Measure the voltage V_4 and compare with the result of part 4(b).

V_4(measured) = _____

Superposition Principle (dc)

OBJECT

To verify experimentally the superposition principle as applied to dc circuits.

EQUIPMENT REQUIRED

Resistors

1 — 1.2-kΩ, 2.2-kΩ, 3.3-kΩ

Instruments

1 — DMM (or VOM)

2* — dc Power supplies

*The unavailability of two power supplies may require that two groups work together.

EQUIPMENT ISSUED

TABLE 9.1

Item	Manufacturer and Model No.	Laboratory Serial No.
DMM (or VOM)		
Power supply		
Power supply		

TABLE 9.2

Resistors	
Nominal Value	Measured Value
1.2 kΩ	
2.2 kΩ	
3.3 kΩ	

RÉSUMÉ OF THEORY

The superposition principle states that the current through, or voltage across, any resistive branch of a multisource network is the algebraic sum of the contributions due to each source acting independently. When the effects of one source are considered, the others are replaced by their internal resistances. Superposition is effective only for linear circuit relationships.

This principle permits one to analyze circuits without resorting to simultaneous equations. Nonlinear effects, such as power, which varies as the square of the current or voltage, cannot be analyzed using this principle.

PROCEDURE

Part 1   Calculated Results

Insert the measured resistor values in Fig. 9.1.

$R_1 = 1.2 \text{ k}\Omega$ $R_3 = 3.3 \text{ k}\Omega$

$R_{1 \text{ measured}} = \underline{\hspace{2cm}}$
$R_{2 \text{ measured}} = \underline{\hspace{2cm}}$
$R_{3 \text{ measured}} = \underline{\hspace{2cm}}$

Caution: Be sure dc supplies have common ground.

FIG. 9.1

(a) Calculate the currents I_1, I_2, and I_3 for the source E_1 using the diagram of Fig. 9.2 and the measured resistor values.

FIG. 9.2

$I_1 = $ _____

$I_2 = $ _____

$I_3 = $ _____

(b) Calculate the currents I_1, I_2, and I_3 for the source E_2 using the diagram of Fig. 9.3 and the measured resistor values.

FIG. 9.3

$I_1 = $ _____

$I_2 = $ _____

$I_3 = $ _____

(c) Determine I_1, I_2, and I_3 for the network of Fig. 9.1 using the calculated results of parts 1(a) and 1(b). Show their resulting directions on Fig. 9.1.

$I_1 = $ _____

$I_2 = $ _____

$I_3 = $ _____

(d) Using the results of part 1(a), calculate the voltage across each resistor using Ohm's Law. Insert the polarities of V_1, V_2, and V_3 on Fig. 9.2.

$V_1 = $ _____

$V_2 = $ _____

$V_3 = $ _____

(e) Using the results of part 1(b), calculate the voltage across each resistor using Ohm's Law. Insert the polarities of V_1, V_2, and V_3 on Fig. 9.3.

$V_1 = $ _____

$V_2 = $ _____

$V_3 = $ _____

(f) Determine V_1, V_2, and V_3 for the network of Fig. 9.1 using the calculated results of parts 1(d) and 1(e). Insert their polarities on Fig. 9.1.

(g) Using the results of part 1(a), calculate the power to each resistor.

$P_1 = $ _____

$P_2 = $ _____

$P_3 = $ _____

(h) Using the results of part 1(b), calculate the power to each resistor.

$P_1 = $ _____

$P_2 = $ _____

$P_3 = $ _____

(i) Using the results of part 1(c), calculate the power to each resistor.

$P_1 = $ _____

$P_2 = $ _____

$P_3 = $ _____

(j) Add the power levels determined in parts 1(g) and 1(h) and compare with the levels determined in part 1(i).

$P_1 \, [1(g)] + P_1 \, [1(h)] = $ _____

$P_1 \, [1(i)] = $ _____

$P_2 \, [1(g)] + P_2 \, [1(h)] = $ _____

$P_2 \, [1(i)] = $ _____

$$P_3 \ [1(g)] + P_3 \ [1(h)] = \text{_____}$$

$$P_3 \ [1(i)] = \text{_____}$$

How do the results compare? Why?

Part 2 Measured Values

(a) Construct the network of Fig. 9.2 and measure the voltages V_1, V_2, and V_3.

$V_1 = \text{_____}$

$V_2 = \text{_____}$

$V_3 = \text{_____}$

How do they compare with the results of part 1(d)?

(b) Construct the network of Fig. 9.3 and measure the voltages V_1, V_2, and V_3.

$V_1 = \text{_____}$

$V_2 = \text{_____}$

$V_3 = \text{_____}$

How do they compare with the results of part 1(e)?

(c) Construct the network of Fig. 9.1 and measure the voltages V_1, V_2, and V_3.

$V_1 = \text{_____}$

$V_2 = \text{_____}$

$V_3 = \text{_____}$

How do they compare with the results of part 1(f)?

(d) Calculate the currents I_1, I_2, and I_3 from the measured values of part 2(a).

$I_1 =$ _____

$I_2 =$ _____

$I_3 =$ _____

How do they compare with the results of part 1(a)?

(e) Calculate the currents I_1, I_2, and I_3 from the measured values of part 2(b).

$I_1 =$ _____

$I_2 =$ _____

$I_3 =$ _____

How do they compare with the results of part 1(b)?

(f) Calculate the currents I_1, I_2, and I_3 from the results of part 2(c).

$I_1 =$ _____

$I_2 =$ _____

$I_3 =$ _____

How do they compare with the results of part 1(c)?

(g) Using the results of part 2(f), calculate the power to each resistor.

$P_1 =$ _____

$P_2 =$ _____

$P_3 =$ _____

How do they compare with the results of part 1(i)?

Part 3 A Second Configuration

(a) Construct the network of Fig. 9.4. Insert the measured resistor values.

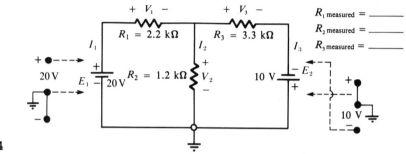

R_1 measured = _____

R_2 measured = _____

R_3 measured = _____

FIG. 9.4

(b) Using the measured resistor values, calculate the voltages V_1, V_2, and V_3 using superposition. Show your work in the space provided.

$V_1 =$ _____ , $V_2 =$ _____ , $V_3 =$ _____

(c) Energize the network of Fig. 9.4 and measure the voltages V_1, V_2, and V_3. Is the superposition principle verified?

$V_1 =$ _____ , $V_2 =$ _____ , $V_3 =$ _____

(d) Using measured resistor values, calculate the voltage V_2 due only to source #1 (E_1). Modify the network and measure the voltage V_2 due only to source #1 (E_1). Be sure to indicate the polarity of V_2.

V_2(calculated) = _____ , V_2(measured) = _____

(e) Repeat (d) for source #2 (E_2).

V_2(calculated) = _____ , V_2(measured) = _____

(f) Using the measured results of (d) and (e), determine the voltage V_2 and compare to the result of (c). Calculate the percent difference using the following equation:

$$\% \text{ Difference} = \frac{|V_2[\text{from (f)}] - V_2[\text{from (c)}]|}{V_2[\text{from (c)}]} \times 100\%$$

V_2[from (d) and (e)] = _____ , V_2[from (c)] = _____

% difference = _____

QUESTION

Has the superposition theorem been substantiated by the results of this experiment? Comment accordingly.

Thevenin's Theorem and Maximum Power Transfer

OBJECT

To verify Thevenin's theorem and the maximum power transfer principle.

EQUIPMENT REQUIRED

Resistors

1 — 91-Ω, 220-Ω, 330-Ω, 470-Ω

1 — 0–1-kΩ potentiometer

Instruments

1 — DMM (or VOM)

1 — dc Power supply

EQUIPMENT ISSUED

TABLE 10.1

Item	Manufacturer and Model No.	Laboratory Serial No.
DMM (or VOM)		
Power supply		

TABLE 10.2

Resistors	
Nominal Value	**Measured Value**
91 Ω	
220 Ω	
330 Ω	
470 Ω	

RÉSUMÉ OF THEORY

Through the use of Thevenin's theorem, a complex two-terminal, linear, multisource dc network can be replaced by one having a single source and resistor.

The Thevenin equivalent circuit consists of a single dc source referred to as the *Thevenin voltage* and a single fixed resistor called the *Thevenin resistance*. The Thevenin voltage is the open-circuit voltage across the terminals in question (Fig. 10.1). The Thevenin resistance is the resistance between these terminals with all of the voltage and current sources replaced by their internal resistances (Fig. 10.1).

If a dc voltage source is to deliver maximum power to a resistor, the resistor must have a value equal to the internal resistance of the source. In a complex network, maximum power transfer to a load will occur when the load resistance is equal to the

FIG. 10.1

Equivalent Circuit

Thevenin resistance "seen" by the load. For this value, the voltage across the load will be one-half of the Thevenin voltage. In equation form, the maximum power is given by

$$P_{\max} = \frac{E_{Th}^2}{4R_{Th}}$$

(10.1)

PROCEDURE

Part 1 Thevenin's Theorem

Calculations:

(a) Insert the measured resistor values into Fig. 10.2 and *calculate* the Thevenin voltage and resistance for the network to the left of points *a-a'*.

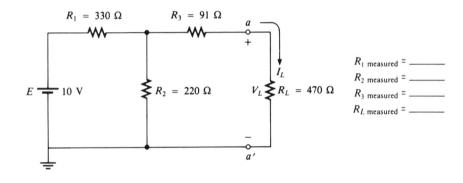

FIG. 10.2

$E_{Th} =$ _____ , $R_{Th} =$ _____ . Enter these values in column 1 of Table 10.3.

(b) Insert the values of E_{Th} and R_{Th} in Fig. 10.3 and calculate I_L.

FIG. 10.3

$I_L =$ _____ . Enter this value in the first space in column 3 of Table 10.3.

(c) Calculate the current I_L in the original network of Fig. 10.2 using series-parallel techniques (use measured resistor values).

$I_L =$ _____ . Enter this value in the second space in column 3 of Table 10.3.

Measurements:

(d) Construct the network of Fig. 10.2 and measure the voltage V_L. Calculate the current I_L. Enter this value in the first space in column 4 of Table 10.3.

V_L(measured) = _____ , I_L(calculated from V_L) = _____

R_{Th}:

(e) Determine R_{Th} by constructing the network of Fig. 10.4 and measuring the resistance between points a-a'.

FIG. 10.4

$R_{Th} = $ _____ . Enter this value in the second space in column 2 of Table 10.3.

E_{Th}:

(f) Determine E_{Th} by constructing the network of Fig. 10.5 and measuring the open-circuit voltage between points a-a'.

FIG. 10.5

E_{Th}(measured) = _____ . Enter this value in the first space in column 2 of Table 10.3.

Thevenin Network:

(g) Construct the network of Fig. 10.6 and set the values obtained for E_{Th} and R_{Th} in parts 1(e) and 1(f), respectively. Use the ohmmeter section of your meter to set the potentiometer properly. Then measure the voltage V_L and calculate the current I_L. Enter this value in the second space in column 4 of Table 10.3.

FIG. 10.6

V_L(measured) = _____ , I_L(calculated from V_1) = _____

TABLE 10.3

Calculated Values of E_{Th} and R_{Th} [Part 1(a)]	Measured Values of E_{Th} and R_{Th} [Parts 1(e) and 1(f)]	Calculated Values of I_L [Parts 1(b) and 1(c)]	Measured Values of I_L [Parts 1(d) and 1(g)]
E_{Th} = _____	E_{Th} = _____ [part 1(f)]	Using Thevenin equivalent: I_L = _____ [part 1(b)]	From Fig. 10.2: I_L = _____ [part 1(d)]
R_{Th} = _____	R_{Th} = _____ [part 1(e)]	Using other methods: I_L = _____ [part 1(c)]	Using Thevenin equivalent: I_L = _____ [part 1(g)]

How do the calculated and measured values of E_{Th} and R_{Th} compare?

How do the calculated and measured values of I_L compare?

Part 2   Maximum Power Transfer

   (a) Construct the network of Fig. 10.7 and set the potentiometer to 50 Ω. Measure the voltage across R as you vary R through the following values: 50, 100, 200, 300, 330, 400, 600, 800, and 1000 Ω. Be sure to set the resistance with the ohmmeter section of

FIG. 10.7

your meter before each reading. (Remember to disconnect the dc supply when setting the resistance level.) Complete Table 10.4 and plot P_R (power) versus R on Graph 10.1.

    (b) Theoretically, for what value of R will the power delivered to R be maximum? Check this value against that obtained from the graph.

$R =$ _____

    In addition, what should the voltage across R be when R is set for maximum power? Do your experimental data substantiate this?

$V_R =$ _____

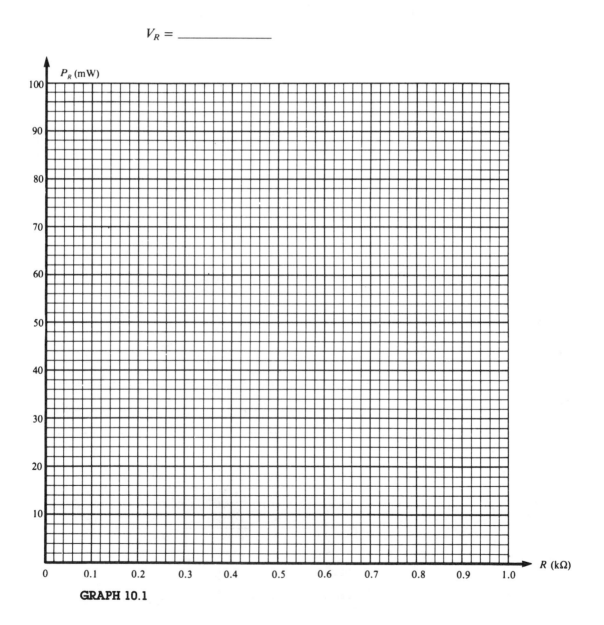

GRAPH 10.1

TABLE 10.4

R	V_R	$P_R = \dfrac{V_R^2}{R}$
50 Ω		
100 Ω		
200 Ω		
300 Ω		
R_1 (measured)		
400 Ω		
600 Ω		
800 Ω		
1000 Ω		

(c) Construct the network of Fig. 10.8 and set the potentiometer to 10 Ω. Complete Table 10.5 for each value of R.

FIG. 10.8

TABLE 10.5

R	V_R	P_R
10 Ω		
50 Ω		
100 Ω		
150 Ω		
200 Ω		
250 Ω		
300 Ω		
350 Ω		
400 Ω		

(d) What value of R would appear to result in a maximum power transfer to the load? How does it compare to the value of R_{Th} determined in Part 1(a)?

$R =$ _____

Part 3   Concluding Exercise

(a) Construct the network of Fig. 10.9. Insert the measured resistor values.

$R_{1\,\text{measured}} =$ _____
$R_{2\,\text{measured}} =$ _____
$R_{3\,\text{measured}} =$ _____

FIG. 10.9

(b) Using *experimental techniques*, determine the maximum power to the resistor R_L. That is, determine E_{Th} and R_{Th} experimentally and then use Eq. (10.1).

$E_{Th} = $ _____ , $R_{Th} = $ _____ , $P_{max} = $ _____

(c) Calculate the values of E_{Th}, R_{Th}, and P_{max} using measured resistor values and compare to the results of (b), using Table 10.6.

TABLE 10.6

Parameter	(b)	(c)	% Difference = $\dfrac{\|(c) - (b)\|}{(b)} \times 100\%$
E_{Th}			
R_{Th}			
P_{max}			

(d) Comment on whether $R_L = 91\,\Omega$ is a poor or excellent choice for maximum power transfer to the load. With $R_L = 91\,\Omega$, what percent of maximum power is delivered to the load using the equation $(P_{91\Omega}/P_{max}) \times 100\%$?

% of maximum power = _____

QUESTION

Is Thevenin's Theorem verified by the results of this experiment? Be specific.

Norton's Theorem and Source Conversions

OBJECT

To verify Norton's theorem and the theory of source conversion.

EQUIPMENT REQUIRED

Resistors

1 — 47-Ω, 220-Ω, 330-Ω, 3.3-kΩ, 10-kΩ

1 — 0–1-kΩ potentiometer

Instruments

1 — DMM

1 — dc Power supply

EQUIPMENT ISSUED

TABLE 11.1

Item	Manufacturer and Model No.	Laboratory Serial No.
DMM		
Power supply		

TABLE 11.2

Resistors	
Nominal Value	**Measured Value**
47 Ω	
220 Ω	
330 Ω	
3.3 kΩ	
10 kΩ	

RÉSUMÉ OF THEORY

Through the use of Thevenin's or Norton's theorem, a complex two-terminal, linear, multisource dc network can be replaced by a single source and resistor.

The Norton equivalent circuit is a single dc current source in parallel with a resistor. The *Norton current* is the short-circuit current between the terminals in question. The *Norton resistance* is the resistance between these terminals with all voltage and current sources replaced by their internal resistances.

The theory of source conversion dictates that the Norton and Thevenin circuits be terminally equivalent and related as follows:

$$\boxed{R_N = R_{Th}} \qquad \boxed{E_{Th} = I_N R_N} \qquad \boxed{I_N = \frac{E_{Th}}{R_{Th}}} \qquad \textbf{(11.1)}$$

Therefore, if you have converted a complex network to its Norton equivalent, it is a simple matter to convert it to a Thevenin equivalent circuit.

PROCEDURE

Part 1   Norton Equivalent Circuit

(a) Construct the circuit of Fig. 11.1. Insert the measured values of each resistor.

FIG. 11.1

Measure the current through the 47-Ω resistor by measuring the voltage drop across the resistor (V_{ab}) and then calculating the value of I using Ohm's law (use the measured resistor value). Record the value of I.

$V_{ab} =$ _____ , $I =$ _____

In the parts that follow, we will show that this same current will result when we substitute the Norton equivalent circuit.

(b) Calculate the Norton resistance (R_N) and the Norton current (I_N) for the network external to the 47-Ω resistor. Use the measured values of the resistors for your calculations. (Show all calculations and be neat.)

Record the calculated values of R_N and I_N in Table 11.3.

TABLE 11.3

Parameter	Calculated	Measured	% Difference
R_N			
I_N			

(c) Remove the 47-Ω resistor and initially place the DMM set on a high current range across points *a-b*. Turn on the power and measure the current I_{a-b} with the scale setting that gives you the most accurate reading. This is the value of I_N and should be recorded in Table 11.3. Shut down, physically disconnect the supply from the circuit, and connect a wire across points *c-d*. Change the DMM mode so that it is operating as an ohmmeter and measure the resistance across points *a-b*. This is the measured value of R_N. Record this value in Table 11.3. How do the two values I_N and R_N compare? Calculate the percent difference using

$$\% \text{ Difference} = \frac{|\text{measured} - \text{calculated}|}{\text{calculated}} \times 100\%$$

% difference = _____

Draw the Norton equivalent circuit in the space below. Indicate the points *a-b*, the 47-Ω resistor, and the values of R_N and I_N on the diagram. Use the measured values of R_N and I_N.

(d) Construct the circuit of Fig. 11.2.

FIG. 11.2

Vary the supply voltage until the DMM indicates the value I_N from (c). Record the values of E and I_N in Fig. 11.3(b). Next remove the DMM and, using it as an ohmmeter, set the 0–1-kΩ potentiometer to the value of R_N from (c). Now insert the 0–1-kΩ potentiometer in the circuit of Fig. 11.3(b).

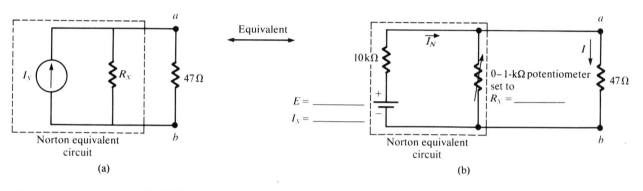

FIG. 11.3

The network of Fig. 11.3(b) is the Norton equivalent circuit. The 0–1-kΩ potentiometer is equivalent to R_N, and the 10-kΩ resistor in series with the power supply is the equivalent current source. The 10-kΩ resistor was chosen to insure minimum sensitivity on the part of I_N to the smaller resistor values connected in parallel in Fig. 11.3(b). In other words, $I_N = E/(10 \text{ k}\Omega + R_{\text{network}}) \cong E/10 \text{ k}\Omega$, and therefore approximates an ideal current source.

Measure the voltage V_{ab} and compute I using the measured resistor value.

$V_{ab} = $ _____ , $I = $ _____

How do these values compare to those of (a)? Calculate the percent difference as determined by

$$\% \text{ Difference} = \frac{|\text{part (d)} - \text{part (a)}|}{\text{part (a)}} \times 100\%$$

% difference $(I) = $ _____

Part 2   Source Conversion

(a) According to the theory of source conversion, the Norton equivalent circuit of Fig. 11.3(b) can easily be converted to a Thevenin equivalent circuit. Further, if the circuits are completely equivalent, the values of V_{ab} and I should be the same as before.

The conversion can be accomplished by first calculating the Thevenin voltage E_{Th}, where $E_{Th} = I_N R_N$.

E_{Th} = _____

Since $R_{Th} = R_N$, the Thevenin equivalent circuit can now be constructed.

(b) Construct the circuit of Fig. 11.4

FIG. 11.4   Thevenin equivalent circuit

Now measure the voltage V_{ab} and again compute I. (Use the measured value of the 47-Ω resistor.)

V_{ab} = _____ , I = _____

How do the values of V_{ab} and I compare to those in Part 1(d)? Calculate the percent difference from

$$\% \text{ Difference} = \frac{|\text{part 2(b)} - \text{part 1(d)}|}{\text{part 1(d)}} \times 100\%$$

% difference (I) = _____ , % difference (V_{ab}) = _____

Do the results verify the source conversion theory?

PROBLEMS

Calculate the Norton and Thevenin equivalent circuits for the network of Fig. 11.5 to the left of terminals a-b.

FIG. 11.5

E_{Th} = _____ , R_{Th} = _____ , I_N = _____ ,

R_N = _____

Sketch the Norton and Thevenin equivalent circuits below.

Methods of Analysis

OBJECT

To become familiar with the branch-, mesh-, and nodal-analysis techniques.

EQUIPMENT REQUIRED

Resistors

$1-1.0$-kΩ, 1.2-kΩ

$2-2.2$-kΩ, 3.3-kΩ

Instruments

$1-$DMM (or VOM)

$2*-$dc Power supplies

*The unavailability of two supplies will simply require that two groups work together.

EQUIPMENT ISSUED

TABLE 12.1

Item	Manufacturer and Model No.	Laboratory Serial No.
DMM (or VOM)		
Power supply		
Power supply		

TABLE 12.2

Resistors	
Nominal Value	**Measured Value**
1.0 kΩ	
1.2 kΩ	
2.2 kΩ	
3.3 kΩ	

RÉSUMÉ OF THEORY

The branch-, mesh-, and nodal-analysis techniques are used to solve complex networks with a single source or networks with more than one source that are not in series or parallel.

The branch- and mesh-analysis techniques will determine the currents of the network, while the nodal-analysis approach will provide the potential levels of the nodes of the network with respect to some reference.

The application of each technique follows a sequence of steps, each of which will result in a set of equations with the desired unknowns. It is then only a matter of solving these equations for the various variables, whether they be current or voltage.

PROCEDURE

Part 1   Branch-current Analysis

(a) Construct the network of Fig. 12.1 and insert the measured values of the resistors in the spaces provided.

R_1 measured = _____

R_2 measured = _____

R_3 measured = _____

Caution: Be sure dc supplies are hooked up as shown (common ground) before turning the power on.

FIG. 12.1

(b) Using branch-current analysis, *calculate* the current through each branch of the network of Fig. 12.1 and insert in Table 12.3. Use the measured resistor values and assume the current directions shown in the figure.

TABLE 12.3

Current	Calculated	Measured
I_1		
I_2		
I_3		

(c) Measure the voltages V_1, V_2, and V_3 and enter below with a minus sign for any polarity that is opposite to that in Fig. 12.1.

$V_1 = $ _____ , $V_2 = $ _____ , $V_3 = $ _____

Calculate the currents I_1, I_2, and I_3 and insert in Table 12.3 as the measured values. Be sure to include a minus sign if the current direction is opposite to that appearing in Fig. 12.1.

How do the calculated and measured results compare?

Part 2   Mesh Analysis

    (a) Construct the network of Fig. 12.2 and insert the measured values of the resistors in the spaces provided.

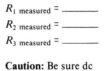

R_1 measured = _____

R_2 measured = _____

R_3 measured = _____

Caution: Be sure dc supplies are hooked up as shown (common ground) before turning the power on.

FIG. 12.2

    (b) Using mesh analysis, *calculate* the mesh currents of the network. Use the measured resistor values and the indicated directions for the mesh currents. Then determine the current through each resistor and insert in Table 12.4 in the "Calculated" column.

Mesh currents: $I_1 = $ _____ , $I_2 = $ _____

TABLE 12.4

Current	Calculated	Measured
I_{R1}		
I_{R2}		
I_{R3}		

(c) Measure the voltages V_1, V_2, and V_3 and enter below with a minus sign for any polarity that is opposite to that in Fig. 12.2.

$V_1 =$ _____ , $V_2 =$ _____ , $V_3 =$ _____

Calculate the currents I_{R_1}, I_{R_2}, and I_{R_3} and insert in Table 12.4 as the measured values. Be sure to include a minus sign if the current direction is opposite to that defined by the polarity of the voltage across the resistor. $I_{R_1} = I_1$, $I_{R_2} = I_2$, and $I_{R_3} = I_1 - I_2$, as defined by Fig. 12.2.

How do the calculated and measured results compare?

Part 3   Nodal Analysis

(a) Construct the network of Fig. 12.3 and insert the measured resistor values.

FIG. 12.3

R_1 measured = _____
R_2 measured = _____
R_3 measured = _____
R_4 measured = _____

Caution: Be sure dc supplies are hooked up as shown (common ground) before turning the power on.

Calculations:

(b) Convert both voltage sources to current sources.

(c) Determine V_a using nodal analysis.

V_a(calculated) = _____

(d) Using V_a, calculate the values of I_{R_1}, I_{R_2}, I_{R_3}, and I_{R_4} and insert in Table 12.5.

TABLE 12.5

Current	Calculated	Measured
I_{R1}		
I_{R2}		
I_{R3}		
I_{R4}		

Measurements:

(e) Energize the network and measure the voltage V_a. Compare with the result of part 3(c).

V_a(measured) = _____

(f) Calculate the currents I_{R_1}, I_{R_2}, I_{R_3}, and I_{R_4} and insert in Table 12.5 as the measured results.

(g) How do the calculated and measured results compare?

Part 4 Bridge Network

(a) Construct the network of Fig. 12.4. Insert the measured resistor values.

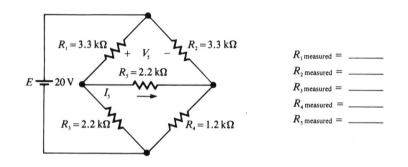

FIG. 12.4

(b) Using any one of the three techniques examined in this experiment, calculate the voltage V_5 and the current I_5. Use the measured resistor values.

V_5(calculated) = _____ , I_5(calculated) = _____

(c) Measure the volage V_5 and insert below with a minus sign if the polarity is different from that appearing in Fig. 12.4.

V_5(measured) = _____

(d) Calculate the percent difference between the two values.

% difference = _____

QUESTION

Many times one is faced with the question of which method to use in a particular problem. The laboratory activity does not prepare one to make such choices but only shows that the methods work and are solid. From your experience in this activity, summarize in your own words which method you would choose and why.

Capacitors

OBJECT

To become familiar with the characteristics of a capacitor in a dc system.

EQUIPMENT REQUIRED

Resistors

2 — 1.2-kΩ, 100-kΩ

Capacitors

2 — 100-μF (electrolytic)

Instruments

1 — DMM (or VOM)

1 — dc Power supply

1 — Single-pole, single-throw switch

EQUIPMENT ISSUED

TABLE 13.1

Item	Manufacturer and Model No.	Laboratory Serial No.
DMM (or VOM)		
Power supply		

TABLE 13.2

Resistors	
Nominal Value	**Measured Value**
1.2 kΩ	
1.2 kΩ	
100 kΩ	
100 kΩ	

RÉSUMÉ OF THEORY

The resistor dissipates electrical energy in the form of heat. In contrast, the capacitor is a component that stores electrical energy in the form of an electrical field. The capacitance of a capacitor is a function of its geometry and the dielectric used. Dielectric materials (insulators) are rated by their ability to support an electric field in terms of a figure called the dielectric constant (k). Dry air is the standard dielectric for purposes of reference and is assigned the value of unity. Mica, with a dielectric constant of 6, has six times the capacitance of a similarly constructed air capacitor.

One of the most common capacitors is the parallel-plate type. The capacitance (in farads) is determined by

$$C = 8.85 \times 10^{-12} \epsilon_r \frac{A}{d} \quad \text{(farads, F)} \tag{13.1}$$

where ϵ_r is the relative permittivity (or dielectric constant), A is the area of the plates (m^2), and d is the distance between the plates (m). By changing any one of the three parameters, one can easily change the capacitance.

The dielectric constant is not to be confused with the dielectric strength of a material, which is given in volts per unit length and is a measure of the maximum stress that a dielectric can withstand before it breaks down and loses its insulator characteristics.

In a dc circuit, the volt-ampere characteristics of a capacitor in the steady-state mode are such that the capacitor prevents the flow of dc current but will charge up to a dc voltage. Essentially, therefore, the characteristics of a capacitor in the steady-state mode are those of an *open circuit*. The charge Q stored by a capacitor is given by

$$Q = CV \qquad\qquad\text{(13.2)}$$

where C is the capacitance and V is the voltage impressed across the capacitor.
Capacitors in series behave like resistors in parallel:

$$\frac{1}{C_T} = \frac{1}{C_1} + \frac{1}{C_2} + \frac{1}{C_3} + \cdots + \frac{1}{C_N} \qquad\qquad\text{(13.3)}$$

Capacitors in parallel behave like resistors in series:

$$C_T = C_1 + C_2 + C_3 + \cdots + C_N \qquad\qquad\text{(13.4)}$$

The energy stored by a capacitor (in joules) is determined by

$$W = \frac{1}{2} CV^2 \qquad\qquad\text{(13.5)}$$

For the network of Fig. 13.1, the capacitor will, for all practical purposes, charge up to E volts in *five* time constants, where a time constant (τ) is defined by

$$\tau = RC \qquad\qquad\text{(13.6)}$$

FIG. 13.1

In one time constant, the voltage v_C will charge up to 63.2% of its final value, in 2τ up to 86.5%, in 3τ up to 95.1%, in 4τ up to 98.1%, and in 5τ up to 99.3%, as defined by

$$v_C = E(1 - e^{-t/RC}) \qquad\qquad\text{(13.7)}$$

The current i_C is defined by

$$i_C = \frac{E}{R} (e^{-t/RC}) \qquad\qquad\text{(13.8)}$$

PROCEDURE

Part 1

(a) Construct the network of Fig. 13.2. Insert the measured resistor value. Be sure to note polarity on electrolytic capacitors as shown in the figure.

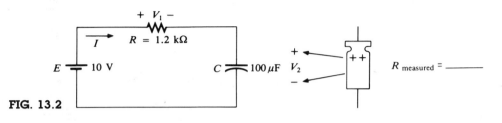

FIG. 13.2

(b) Calculate the steady-state value (defined by a period of time greater than five time constants) of the current I and the voltages V_1 and V_2.

$I =$ _____ , $V_1 =$ _____ , $V_2 =$ _____

(c) Measure the voltages V_1 and V_2 and calculate the current I from Ohm's law. Compare with the results of part 1(b) above.

$I =$ _____ , $V_1 =$ _____ , $V_2 =$ _____

(d) Calculate the energy stored by the capacitor.

$W =$ _____

(e) Carefully disconnect the supply and measure the voltage across the discon-
nected capacitor. Is there a reading? Why?

$V_C =$ _____

(f) Short the capacitor terminals with a lead. Why was it necessary to perform
this step?

Part 2

(a) Construct the network of Fig. 13.3. Insert the measured resistor values.

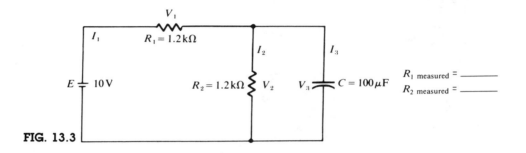

R_1 measured = _____

R_2 measured = _____

FIG. 13.3

(b) Using ideal elements, calculate the theoretical steady-state levels (time
greater than five time constants) of the following quantities. (The measured resistor values should be used.)

$I_1 =$ _____ , $I_2 =$ _____ , $I_3 =$ _____

$V_1 =$ _____ , $V_2 =$ _____ , $V_3 =$ _____

(c) Energize the system and measure the voltages V_1, V_2, and V_3. Calculate the
currents I_1 and I_2 from Ohm's Law and the current I_3 from Kirchhoff's current law. Compare the results with those of part 2(b).

$V_1 =$ _____ , $V_2 =$ _____ , $V_3 =$ _____

$I_1 =$ _____ , $I_2 =$ _____ , $I_3 =$ _____

Part 3

(a) Construct the network of Fig. 13.4. Insert the measured resistor values.

R_1 measured = _____

R_2 measured = _____

FIG. 13.4

(b) Assuming ideal elements, and using measured resistor values, calculate the theoretical steady-state levels of the following quantities:

$I_1 =$ _____ , $I_2 =$ _____ , $I_3 =$ _____ ,

$I_4 =$ _____

$V_1 =$ _____ , $V_2 =$ _____ , $V_3 =$ _____ ,

$V_4 =$ _____

(c) Energize the system and measure the voltages V_1, V_2, V_3, and V_4. Compare the results with those in part 3(b).

$$V_1 = \underline{\hspace{2cm}} , \quad V_2 = \underline{\hspace{2cm}} , \quad V_3 = \underline{\hspace{2cm}}$$

$$V_4 = \underline{\hspace{2cm}}$$

Part 4 Charging Network

(a) Calculate the time constant for the network of Fig. 13.5. Use the measured resistance value.

FIG. 13.5

$$\tau = \underline{\hspace{3cm}}$$

(b) Discharge the capacitor and make sure the switch is in the open position. Then energize the source, close the switch, and note how many seconds pass before the voltage v_C reaches 63.2% of its final value or $(0.632)(20) = 12.64$ V (one time constant).

$$t_{63.2\%} = \underline{\hspace{3cm}}$$

You will note that the t is somewhat larger than you expected from the calculations of part 4(a). This is due to the leakage resistance typically present in electrolytic capacitances of this capacitance level.

(c) An equivalent capacitance can be defined, however, that will permit an examination of the charging phase of a capacitor when applied to a dc source. It will be determined from

$$\boxed{C_{eq} = \frac{t}{R}} \qquad (13.9)$$

(where t = time required to reach the 63.2% level and R = measured value), as derived from the time constant equation $\tau = RC$.

Calculate the equivalent capacitance for the configuration of Fig. 13.5.

$C_{eq_1} =$ _____

Repeat the above determination for the other capacitor.

$C_{eq_2} =$ _____

From this point on, simply assume that each capacitor has a capacitance equivalent to C_{eq} and the data obtained for this experiment should be quite satisfactory.

(d) Determine the time constant and charging time for the network of Fig. 13.6. Use the measured resistance levels and C_{eq} for each capacitor.

FIG. 13.6

$\tau =$ _____ , $5\tau =$ _____

(e) Using a watch, record (to the best of your ability) the voltage across the resistors at the time intervals noted in Table 13.3 after the switch is closed. Complete the table for V_C using the fact that $V_C = E - V_R$.

TABLE 13.3

t (s)	5	10	15	20	25	30	35	40	45	50
V_R										
V_C										

    (f) In one time constant, the voltage V_C should be 63.2% of its final steady-state value of 20 V. Is this verified by the data of part 4(e)?

    (g) On Graph 13.1, plot the curves of v_R and v_C versus time.

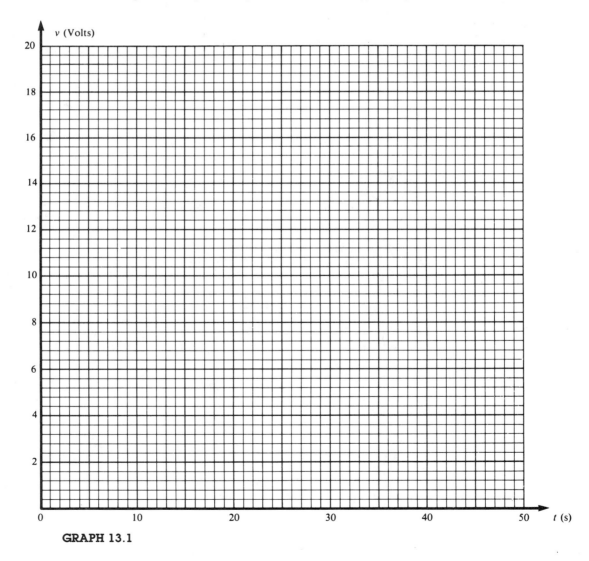

GRAPH 13.1

(h) Construct the network of Fig. 13.7. Insert the measured resistance level and each C_{eq}.

Repeat parts 4(d) through 4(g) for Fig. 13.7 and complete Table 13.4.

FIG. 13.7

$\tau =$ _____ , $5\tau =$ _____

TABLE 13.4

t (s)	10	20	30	40	50	60	70	80	90	100
V_R										
V_C										

On Graph 13.2, plot the curves of v_R and v_C versus time. Do the data verify the theoretical conclusions?

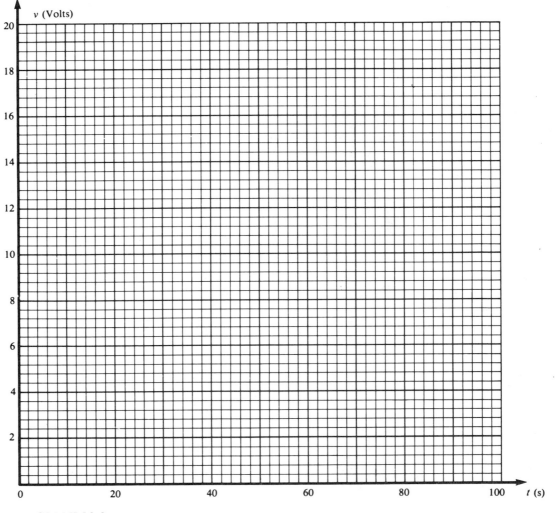

GRAPH 13.2

R-L and *R-L-C* Circuits with a dc Source Voltage

OBJECT

To investigate the response of R-L and R-L-C circuits to a dc voltage input.

EQUIPMENT REQUIRED

Resistors

1 — 10-Ω, 470-Ω, 1-kΩ, 2.2-kΩ

Inductors

1 — 10-mH

Capacitors

1 — 1-μF

Instruments

1 — DMM

1 — dc Power supply

Miscellaneous

1 — SPST switch

1 — NE-2 neon glow lamp

EQUIPMENT ISSUED

TABLE 14.1

Item	Manufacturer and Model No.	Laboratory Serial No.
DMM		
Power supply		

TABLE 14.2

Resistors	
Nominal Value	Measured Value
10 Ω	
470 Ω	
1 kΩ	
2.2 kΩ	

RÉSUMÉ OF THEORY

The inductor, like the capacitor, is an energy-storing device. The capacitor stores energy in the form of an electric field while the inductor stores it in the form of a magnetic field. The energy stored by an inductor (in joules) is given by

$$W = \frac{1}{2} LI^2$$

(14.1)

In any circuit containing an inductor, the voltage across the inductor is determined by the inductance (L) and the rate of change of the current through the inductor:

$$v_L = L \,\frac{\text{change of current}}{\text{change of time}} = L \,\frac{\Delta I}{\Delta t}$$

(14.2)

If the current is constant (as in a dc circuit), $\Delta I/\Delta t = 0$, and the only voltage drop across the inductor is due to the dc resistance of the wire that makes up the inductor. Because an inductor has a dc resistance, we draw it schematically as shown in Fig. 14.1.

FIG. 14.1

The total inductance of inductors in series is given by

$$L_T = L_1 + L_2 + L_3 + \cdots + L_N \tag{14.3}$$

and in parallel by

$$\frac{1}{L_T} = \frac{1}{L_1} + \frac{1}{L_2} + \frac{1}{L_3} + \cdots + \frac{1}{L_N} \tag{14.4}$$

Note the correspondence of Eqs. (14.3) and (14.4) to the equations for resistive elements.

PROCEDURE

Part 1

Construct the circuit of Fig. 14.2. Insert the measured values of R_1 and R_l.

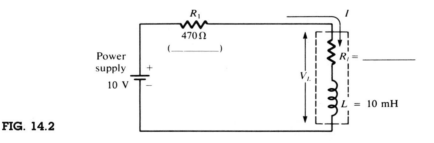

FIG. 14.2

(a) Calculate the current (I) and the voltage (V_L) for steady-state conditions. Since $5\tau = 5L/R = 5(10 \times 10^{-3})/470 \cong 0.1$ ms, steady-state conditions essentially exist once the network is constructed.

$I = $ _____ , $V_L = $ _____

(b) Measure V_L and determine I. (I is to be calculated by measuring the voltage across the 470-Ω resistor and using Ohm's Law.)

$I = $ _____ , $V_L = $ _____

Calculate R_l.

$R_l = $ _____

How does the value of R_l in part 1(b) compare with the measured value?

Is it possible for a practical inductor to have $R_l = 0$? Explain.

Part 2   Parallel R-L dc Circuit

Construct the circuit of Fig. 14.3. Insert the measured value of each resistor.

FIG. 14.3

(a) Calculate I, I_1, and I_2 assuming an ideal inductor $(R_l = 0 \ \Omega)$. Use measured resistor values.

$I = $ _____ , $I_1 = $ _____ , $I_2 = $ _____

(b) Measure V_1 and V_2, and determine I, I_1, and I_2 using the equations

$$I = \frac{V_2}{R_2} \qquad I_1 = \frac{V_1}{R_1} \qquad I_2 = I - I_1$$

$V_1 = $ _____ , $V_2 = $ _____

$I = $ _____ , $I_1 = $ _____ , $I_2 = $ _____

How do the calculated and measured values for I, I_1, and I_2 compare?

Part 3   Series-Parallel *R-L-C* dc Circuit

Construct the network of Fig. 14.4. Insert the measured values of all resistors (including R_l).

FIG. 14.4

(a) Using measured values, calculate I, I_2, and V_C.

$I = $ _____ , $I_2 = $ _____ , $V_C = $ _____

(b) Measure V_1 and V_C and calculate I and I_2 from the equations

$$I = V_1/R_1 \qquad I_2 = \frac{V_2}{R_2} = \frac{V_C}{R_2}$$

Compare the results with those in part 3(a).

$V_1 = $ _____ , $V_C = $ _____

$I = $ _____ , $I_2 = $ _____

(c) Calculate the energy stored by the capacitor and inductor.

$$W_C = \underline{\hspace{3cm}} \, , \, W_L = \underline{\hspace{3cm}}$$

Part 4   Induced Voltage in an Inductor

(a) Place your power supply in series with the supply of another squad as shown in Fig. 14.5

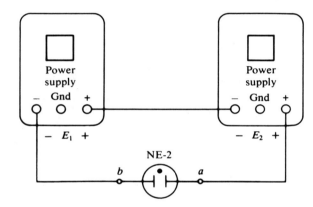

FIG. 14.5

Vary the voltages of both supplies until the lamp lights. Measure the voltage V_{ab} required to light the lamp, noting that $V_{ab} = E_1 + E_2$.

$$V_{ab} = \underline{\hspace{3cm}}$$

Repeat the above for the lamp of the other squad.

(b) Return the other supply and construct the circuit of Fig. 14.6.

FIG. 14.6

Set the voltage source E to zero volts. Then close the switch and increase E until the voltage $V_L = 0.3$ V [significantly less than V_{ab} from part 4(a)]. Measure the voltage V_1 and calculate the current I delivered by the source to establish the magnetic field of the coil. Assume the bulb current is zero amperes.

$V_1 =$ _____ , $I =$ _____

Calculate the energy stored by the coil using $W_L = 1/2LI^2$.

$W_L =$ _____

Does the lamp light with the switch in the on (closed) position? Why not?

Now open the switch and note the effect on the lamp. Describe what happens.

Since $v_L = L(di/dt)$, the change in current di/dt was sufficiently high to develop a product $L(di/dt)$ that would equal the voltage V_{ab} measured earlier. Since the current dropped from the level I determined earlier to zero, $\Delta I = I - 0 = I$ in Eq. (14.2) and V_L must be at least V_{ab} volts. Substituting, we obtain

$$v_L = L \frac{\Delta I}{\Delta t}$$

$$V_{ab} = \frac{LI}{\Delta t}$$

and the switching time Δt can be determined from

$$\Delta t = \frac{LI}{V_{ab}}$$

Calculate Δt for your level of I and V_{ab}.

$\Delta t =$ _____

What is the effect on Δt of making the inductance larger? Comment accordingly.

An automobile ignition coil (an inductor) has an inductance of 5 H. The initial current is 2 A, and $\Delta t = 0.4$ ms. Calculate the induced voltage at the spark plug.

$V_L = $ _____

Design of a dc Ammeter and Voltmeter and Meter Loading Effects

OBJECT

To design and calibrate a dc ammeter and voltmeter.

EQUIPMENT REQUIRED

Resistors

1 — 10-Ω, 47-Ω

2 — 100-kΩ

1 — 0–10-kΩ potentiometer

Instruments

1 — VOM

1 — DMM

1 — dc Power supply

1 — 1-mA, 1000-Ω d'Arsonval meter movement

EQUIPMENT ISSUED

TABLE 15.1

Item	Manufacturer and Model No.	Laboratory Serial No.
VOM		
DMM		
Power supply		
Meter movement		

TABLE 15.2

Resistors	
Nominal Value	**Measured Value**
10 Ω	
47 Ω	
100 kΩ	
100 kΩ	

RÉSUMÉ OF THEORY

The basic meter movement is a current-sensitive device. Its construction consists of a coil suspended in a magnetic field with a pointer attached to it. When current is passed through the coil, an interaction of magnetic fields will cause the coil to rotate on its axis. The attached pointer will then indicate a particular angular displacement proportional to the magnitude of the current through the coil. D'Arsonval meter movements are delicate and sensitive dc milliammeters or microammeters used in the design and construction of dc voltmeters and ammeters.

A dc movement (d'Arsonval) can be used to construct a dc voltmeter by connecting a resistor in series with the movement, as shown in Fig. 15.1(a). This series resistor is called the *multiplier*. An ammeter (Fig. 15.1b) is constructed by placing a resistor in parallel with the meter movement. This resistor is called the *shunt* resistor.

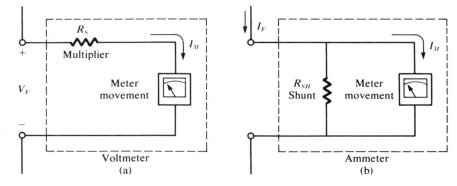

FIG. 15.1

The value of R_S can be calculated as follows:

$$R_S = \frac{V_F}{I_M} - R_M$$

The value of R_{SH} can be calculated as follows:

$$R_{SH} = \left(\frac{I_M}{I_F - I_M}\right)(R_M) \quad \textbf{(15.1)}$$

where: R_S = multiplier resistor

R_{SH} = shunt resistor

I_M = full-scale deflection current of the meter movement

R_M = internal resistance of the meter movement

V_F = maximum value of the voltage to be measured for the range being designed

I_F = maximum value of the current to be measured for the range being designed

EXAMPLE   Using a 0–5-mA meter movement with an internal resistance of 5 kΩ, design

(a) a 0–50-V voltmeter

(b) a 0–100-mA milliammeter

Given $I_M = 5$ mA and $R_M = 5$ kΩ, then

Solutions

(a) $R_S = \dfrac{V_F}{I_M} - R_M = \dfrac{50}{5 \times 10^{-3}} - 5 \times 10^3 = 5 \text{ k}\Omega$

(note Fig. 15.2a) and

(b)

$$R_{SH} = \left(\frac{I_M}{I_F - I_M}\right)(R_M) = \left[\frac{5 \times 10^{-3}}{(100 \times 10^{-3}) - (5 \times 10^{-3})}\right](5 \times 10^3) = 263 \ \Omega$$

(note Fig. 15.2b).

The current sensitivity of a meter is a measure of the current necessary to obtain a full-scale deflection of the meter movement. The inverse of current sensitivity of a voltmeter is the ohm/volt sensitivity of the meter. Therefore, the dc multirange voltmeter, with a 1000-ohm/volt sensitivity, has a current sensitivity of 1/1000 or 1 mA. The VOM, with a 20,000-ohm/volt sensitivity for dc, has a current sensitivity of 1/20,000 or 50 μA.

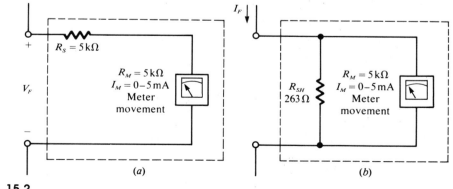

(a)   (b)

FIG. 15.2

Ideally, the internal resistance of a voltmeter should be infinite to minimize its impact on a circuit when taking measurements. As shown in Fig. 15.3, the internal resistance of the meter will appear in parallel with the element under investigation. The effect is to reduce the net resistance of the network since $R_2 > R_2 \| R_{int}$, causing V_2 to decrease in magnitude. The closer R_{int} is to R_2, the more it will affect the reading. As a rule, R_{int} should be more than 10 times R_2 to ensure a reasonable level of accuracy in the reading. The internal resistance of a voltmeter is the series combination of the multiplier and movement resistances, as shown in Fig. 15.3.

The accuracy in percent of the full-scale reading indicates the maximum error that will occur in the meter. The Simpson 377 dc voltmeter has an accuracy of ±5% of full scale. Thus, when set on the 100-V scale, the reading may be in error by ±5 V at any point on the scale. For maximum accuracy, therefore, always choose a scale setting that will give a reading (meter deflection) as close to full scale as possible.

FIG. 15.3

PROCEDURE

Part 1   Design of a dc Voltmeter

(a) Use the 1000-Ω, 1-mA meter movement to design a 0–10-V dc voltmeter using the appropriate variable resistor. Ask your instructor to check your design before proceeding. Show all calculations.

$R_S = $ _____

Calibration of the Designed Voltmeter

(b) Connect your newly constructed voltmeter in parallel with a voltage source and the DMM as shown in Fig. 15.4. Set the variable resistor to R_S with your DMM in the ohmmeter mode.

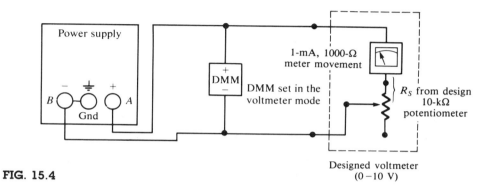

FIG. 15.4

Now vary the power supply from zero to 10 V in 1-V steps and record your readings in Table 15.3.

TABLE 15.3

DMM	Designed Voltmeter
1 V	
2 V	
3 V	
4 V	
5 V	
6 V	
7 V	
8 V	
9 V	
10 V	

Calculate the percent difference of the reading of your voltmeter at three different readings: 1, 5, and 10 V. Use

$$\% \text{ Difference} = \left[\frac{\left| \left(\begin{array}{c} \text{designed voltmeter} \\ \text{reading} \end{array} \right) - \left(\begin{array}{c} \text{DMM} \\ \text{reading} \end{array} \right) \right|}{\text{DMM reading}} \right] \times 100\%$$

% difference (at 1 V) = _____

% difference (at 5 V) = _____

% difference (at 10 V) = _____

What do you conclude from the above results?

What is the sensitivity of your voltmeter?

Part 2   Design of a dc Milliammeter

Use the meter movement (1-mA, 1000-Ω) to design a 100-mA milliammeter. Let your instructor check your design before proceeding. Then connect the milliammeter into the circuit shown in Fig. 15.5 and compare its reading with that of the multirange dc milliammeter set on the proper scale. Start with the power supply set to zero and increase it so that the current I varies from zero to 100 mA in 10-mA steps using the DMM in the milliammeter mode. Record the data in Table 15.4. Show your calculations.

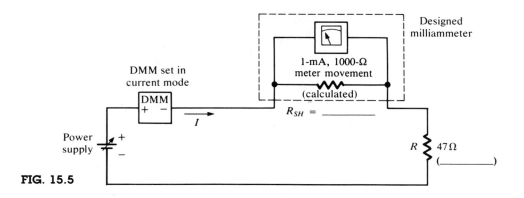

FIG. 15.5

$R_{SH} =$ _____

TABLE 15.4

DMM	Designed Milliammeter
10 mA	
20 mA	
30 mA	
40 mA	
50 mA	
60 mA	
70 mA	
80 mA	
90 mA	
100 mA	

Calculate the percent difference of the reading of your milliammeter at three different readings: 10, 50, and 100 mA. Use the equation provided in part 1.

% difference (at 10 mA) = _____

% difference (at 50 mA) = _____

% difference (at 100 mA) = _____

What conclusions do you draw from the above results?

Part 3   Meter Loading Effects

(a) Construct the circuit of Fig. 15.6.

$R_{1\,measured}$ = _____

$R_{2\,measured}$ = _____

FIG. 15.6

What is the theoretical value of the voltage V_{ab}? Use measured resistor values.

V_{ab} = _____

(b) Read the voltage V_{ab} using the DMM and the VOM on the 10-V scale.

V_{ab} (DMM) = _____ , V_{ab} (VOM) = _____

(c) Most DMMs have an internal impedance of 10 MΩ or greater and can be considered an open circuit across the 100-kΩ resistors. The VOM, however, has an ohm/volt rating of 20,000 and therefore an internal resistance of (10 V)(20,000 Ω/V) = 200 kΩ on the 10-V scale. Using this internal resistance value for the VOM, calculate the voltage V_{ab} and compare with the result of part 3(b) for the VOM.

V_{ab} (calculated − VOM) = _____ ,

V_{ab} [VOM from part 3(b)] = _____

(d) Will the 50-V scale of the VOM give a better reading? Record below and note any disadvantages of using this higher scale.

V_{ab} (50-V scale) = _____

(e) If the resistors were replaced by 1-kΩ values, would there be an improvement in the readings of the VOM? Show why with a sample calculation.

PROBLEMS

1. An ammeter with an accuracy of $\pm 3\%$ of full scale reads 2.1 A when set on the 5-A scale. Within what range of values (about 2.1 A) is this meter accurate?

2. A meter has an ohm/volt sensitivity of 100,000. What is its current sensitivity in microamperes?

Wheatstone Bridge and Δ-Y Conversions

OBJECT

To become familiar with the Wheatstone bridge and Δ-Y conversions.

EQUIPMENT REQUIRED

Resistors

1 — 91-Ω

2 — 220-Ω

3 — 330-Ω, 1-kΩ

1 — 0–1-kΩ potentiometer

1 — Unmarked fixed resistor in the range 47 Ω to 220 Ω

Instruments

1 — DMM (or VOM)

1 — dc Power supply

1 — Commercial Wheatstone bridge (if available)

EQUIPMENT ISSUED

TABLE 16.1

Item	Manufacturer and Model No.	Laboratory Serial No.
DMM (or VOM)		
Power supply		
Wheatstone bridge		

Since more than one resistor of a particular value will appear in some configurations of this experiment, the measured value of each should be determined just prior to inserting them into the network.

RÉSUMÉ OF THEORY

The Wheatstone bridge is an instrument used to make precision measurements of unknown resistance levels. The basic configuration appears in Fig. 16.1. The unknown resistance is R_x, and R_1, R_2, and R_3 are precision resistors of known value.

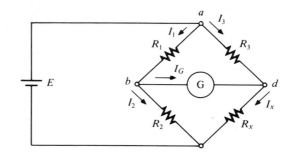

FIG. 16.1

The network is balanced when the galvanometer (G) reads zero.

We know from circuit theory that if $I_G = 0$ A, the voltage V_{bd} is zero, and

$$V_{ab} = V_{ad} \quad \text{and} \quad V_{bc} = V_{dc}$$

By substitution,

$$I_1 R_1 = I_3 R_3 \tag{16.1}$$

and

$$I_2 R_2 = I_x R_x \tag{16.2}$$

Solving Eq. (16.1) for I_1 yields

$$I_1 = \frac{I_3 R_3}{R_1}$$

Substituting I_1 for I_2 and I_3 for I_x in Eq. (16.2) (since $I_1 = I_2$ and $I_3 = I_x$ with $I_G = 0$ A), we have

$$I_1 R_2 = I_3 R_x \quad \text{or} \quad \left(\frac{I_3 R_3}{R_1}\right) R_2 = I_3 R_x$$

Canceling I_3 from both sides and solving for R_x, we obtain

$$R_x = \frac{R_2}{R_1} R_3 \qquad\qquad (16.3)$$

or, in the ratio form,

$$\boxed{\frac{R_1}{R_2} = \frac{R_3}{R_x}} \qquad\qquad (16.4)$$

In the commercial Wheatstone bridge, R_1 and R_2 are variable in decade steps so that the ratio R_1/R_2 is a decimal or integral multiplier. R_3 is a continuous variable resistor, such as a slide-wire rheostat. Before the unknown resistor is connected to the terminals of the commercial bridge, the R_1/R_2 ratio (called the *factor of the ratio arms*) is adjusted for that particular unknown resistor. After the resistor is connected, R_3 is adjusted until there is no detectable current indicated by the galvanometer. (Galvanometer sensitivities are usually 10^{-10} A or better.) The unknown resistance value is the ratio factor times the R_3 setting.

There are certain circuit configurations in which the resistors do not appear to be in series or parallel. Under these conditions, it is necessary to convert the circuit in question from one form to another. The two circuits to be investigated in this experiment are the delta (Δ) and the wye (Y), both of which appear in Fig. 16.2. To convert a Δ to a Y (or vice versa), we use the following conversion equations:

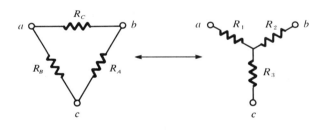

FIG. 16.2

$$R_1 = \frac{R_B R_C}{R_A + R_B + R_C} \qquad R_2 = \frac{R_A R_C}{R_A + R_B + R_C} \qquad R_3 = \frac{R_A R_B}{R_A + R_B + R_C} \qquad (16.5)$$

$$R_A = \frac{R_1 R_2 + R_1 R_3 + R_2 R_3}{R_1} \qquad R_B = \frac{R_1 R_2 + R_1 R_3 + R_2 R_3}{R_2}$$

$$R_C = \frac{R_1 R_2 + R_1 R_3 + R_2 R_3}{R_3} \qquad\qquad (16.6)$$

If $R_A = R_B = R_C$,

$$\boxed{R_Y = \frac{R_\Delta}{3}} \qquad\qquad (16.7)$$

If $R_1 = R_2 = R_3$,

$$\boxed{R_\Delta = 3R_Y}$$

(16.8)

PROCEDURE

Part 1   Wheatstone Bridge Network

(a) Construct the network of Fig. 16.3. Insert the measured value of each resistor and set the potentiometer to the maximum resistance setting.

R_1 measured = _____

R_2 measured = _____

R_3 measured = _____

FIG. 16.3

(b) Starting with the meter on a higher voltage scale, vary the potentiometer until the voltage V_{ab} is as close to zero as possible. Then drop the voltage scales to the lowest range possible to set the voltage V_{ab} as close to zero volts as possible. The bridge is now balanced.

(c) Measure the voltages V_{da}, V_{db}, V_{ac}, and V_{bc}.

$V_{da} = $ _____ , $V_{db} = $ _____ , $V_{ac} = $ _____ ,

$V_{bc} = $ _____

(d) Calculate the currents I_1 and I_3 using Ohm's Law. Are they equal as defined in the Résumé of Theory?

$I_1 = $ _____ , $I_3 = $ _____

(e) Disconnect one lead of the potentiometer (used as a rheostat) and measure its resistance.

$R_{\text{pot}} = R_x = $ _____

Calculate the currents I_2 and I_x using the results of part 1(c) and Ohm's Law. Are they equal as defined in the Résumé of Theory?

$$I_2 = \text{_____} \;, \; I_x = \text{_____}$$

(f) Verify that the following ratio is satisfied:

$$\frac{R_1}{R_3} = \frac{R_2}{R_x}$$

(g) Replace the 91-Ω resistor by the unknown resistor. Proceed as before to adjust the potentiometer until $V_{ab} \cong 0$ V. Remove the variable resistor and measure its resistance with the ohmmeter section of your multimeter.

$$R_{pot} = \text{_____}$$

(h) Calculate the unknown resistance using Eq. (16.4).

$$R_x = \text{_____}$$

(i) What is the maximum value of resistance that this network could measure? Why?

$$R_{max} = \text{_____}$$

Part 2   Commercial Wheatstone Bridge

(a) Use the commercial Wheatstone bridge to measure the resistance of the unknown resistor.

$R_x =$ _____

(b) Measure R_x using the ohmmeter section of your multimeter. Compare with the results of part 2(a) and part 1(h).

$R_x =$ _____

Part 3   Δ-Y Conversions

(a) Construct the network of Fig. 16.4. Insert the measured values of the 220-Ω resistors. Assume for the moment that each 1-kΩ resistor is exactly 1 kΩ.

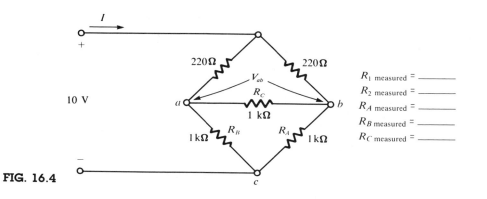

FIG. 16.4

(b) Calculate the current I and the voltage V_{ab} using any method other than a Δ-Y conversion (that is, mesh analysis, nodal analysis, and so on).

$I =$ _____ , $V_{ab} =$ _____

(c) Measure the current I and the voltage V_{ab} and compare to the results of part 3(b).

$I =$ _____ , $V_{ab} =$ _____

(d) Calculate the equivalent Y for the Δ formed by the three 1-kΩ resistors. Draw the equivalent circuit with the Δ replaced by the Y. Insert the values of the resistors in the Y in Fig. 16.5 and also indicate on the diagram those available fixed resistors that have a resistance level closest to the calculated value.

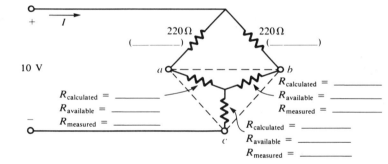

FIG. 16.5

(e) Construct the network drawn in part 3(d) and measure the current I and the voltage V_{ab}. Are they approximately the same as those obtained in part 3(c)? If so, why?

$I =$ _____ , $V_{ab} =$ _____

(f) Calculate the input resistance to the network of part 3(d) using the measured resistor values.

R_T(calculated) = _____

(g) Disconnect the supply and measure the input resistance to the network of (d). Compare with the results of part 3(f).

R_T(measured) = _____

(h) Determine the input resistance to the network of part 3(a) using $R_T = E/I$ and compare with the results of part 3(g). Should they compare? Why?

R_T = _____

(i) Calculate I using the result of part 3(f) and compare with the value measured in part 3(e).

I = _____

Ohmmeter Circuits

OBJECT

To study the various types of circuits used in ohmmeters.

EQUIPMENT REQUIRED

Resistors (fixed)

1 — 1-kΩ, 22-kΩ

2 — 10-kΩ

Resistors (variable)

1 — 0–10-kΩ potentiometer

1 — 0–250-kΩ potentiometer

Instruments

1 — DMM

1 — dc Power supply

1 — 0–1-mA, 1000-Ω meter movement

EQUIPMENT ISSUED

TABLE 17.1

Item	Manufacturer and Model No.	Laboratory Serial No.
DMM		
Power supply		
Meter movement		

TABLE 17.2

Resistors	
Nominal Value	Measured Value
1 kΩ	
10 kΩ	
10 kΩ	
22 kΩ	

RÉSUMÉ OF THEORY

The ohmmeter is an instrument that is used to measure resistance. It is usually only 3% accurate. When resistance is to be measured more accurately, one must use such instruments as the Wheatstone bridge or the Kelvin bridge.

The majority of commercial ohmmeters fall into one of the following three categories:

1. Series type, Fig. 17.1(a)
2. Shunt type, Fig. 17.1(b)
3. Voltage divider type, Fig. 17.1(c)

The variable resistor R_A is a fine-adjust on R_S to establish full-scale deflection (I = rated maximum movement current) with a fixed supply (E).

FIG. 17.1

The *series* type reads right to left. It cannot be used for low-resistance measurements (below 1 Ω).

The *shunt* type reads left to right. It is used primarily for low-resistance measurements (below 1000 Ω).

The *voltage divider* type reads right to left. It is the most versatile and most often used.

In using an ohmmeter, one must remember two important rules:

1. Never leave the ohmmeter on or, in the case of a VOM, in the ohmmeter position. The voltage source is always in the circuit so that the life span of the battery is shortened.
2. Always adjust the "ohms-adjust" after changing the range.

PROCEDURE

Part 1   Series Ohmmeter Circuit

Construct the circuit of Fig. 17.2.

FIG. 17.2

Short the probes points P_1 and P_2 together and adjust the power supply until the meter movement reads 1 mA. Hold the probes apart and read the meter movement. Record in Table 17.3.

For the remaining steps of the experiment, proceed as follows.

Use the DMM to set the potentiometer resistance to the values shown in Table 17.3. Then insert between P_1 and P_2 and read the meter (M). Record in Table 17.3.

Plot the data of Table 17.3 using the vertical and horizontal divisions in Graph 17.1.

TABLE 17.3

Resistance Across Probes	$I_{meter\ movement}$
0 kΩ	1 mA
20 kΩ	
40 kΩ	
60 kΩ	
80 kΩ	
100 kΩ	
120 kΩ	
140 kΩ	
160 kΩ	
180 kΩ	
200 kΩ	
Open circuit	

The graph is now used as a calibration chart for the ohmmeter circuit of Fig. 17.2. That is, for any unknown applied resistor, the resulting current will define an intersection with the plot of Graph 17.1 that will then define the resistance level on the horizontal axis.

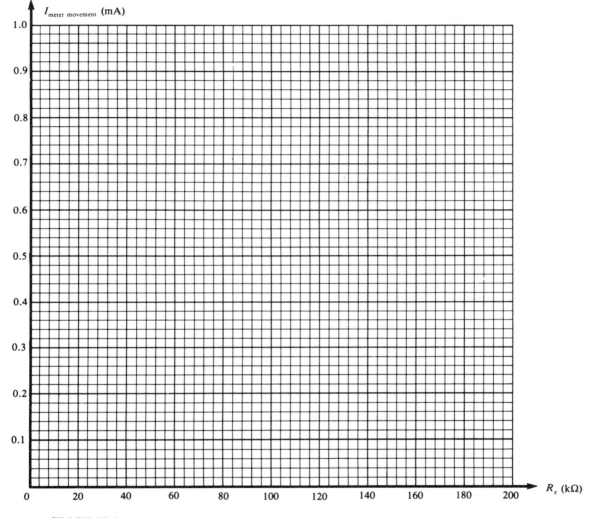

GRAPH 17.1

Vary the 0–250-kΩ potentiometer randomly for three settings and use the designed ohmmeter circuit to determine the unknown resistance. That is, insert the unknown resistance between P_1 and P_2 and note the reading of the meter movement. Then use Graph 17.1 to determine the corresponding resistance level. In each case, use the DMM to verify the results.

$R_1 = $ _____ , $R_2 = $ _____ , $R_3 = $ _____

$R_1(\text{DMM}) = $ _____ , $R_2(\text{DMM}) = $ _____ ,

$R_3(\text{DMM}) = $ _____

Draw a meter scale for your ohmmeter. That is, complete the resistance scale of Fig. 17.3.

FIG. 17.3

Would the scale of this ohmmeter be considered linear or nonlinear?

What is the range of resistance values that this ohmmeter is capable of measuring with reasonably accurate results? Explain.

Why would you recommend that, for the greatest accuracy, the meter be read to the right of the center portion of the scale?

Part 2   Shunt Ohmmeter Circuit

Construct the circuit of Fig. 17.4.

FIG. 17.4

Keeping the probes apart, adjust the power supply until the meter movement reads 1 mA. Record in Table 17.4. Short the probes together; then read the meter. Record in Table 17.4.

Use the DMM to set the potentiometer resistance to the values shown in Table 17.4. Then insert between P_1 and P_2 and read the meter. Record in Table 17.4. Plot the data of Table 17.4 using the horizontal and vertical divisions in Graph 17.2.

TABLE 17.4

Resistance Across Probes	$I_{\text{meter movement}}$
0 Ω	
200 Ω	
400 Ω	
800 Ω	
1,000 Ω	
2,000 Ω	
4,000 Ω	
6,000 Ω	
8,000 Ω	
10,000 Ω	

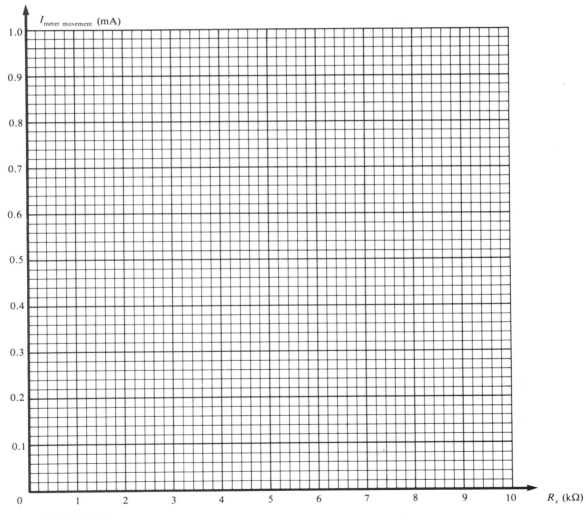

GRAPH 17.2

Graph 17.2 can now be used as a calibration chart for the shunt circuit of Fig. 17.4.

Now set the 0–10-kΩ potentiometer (R_x) to three random values and determine the resistance using the calibration curve. Check the results with the DMM.

$R_1 =$ _____ , $R_2 =$ _____ , $R_3 =$ _____

$R_1(\text{DMM}) =$ _____ , $R_2(\text{DMM}) =$ _____ ,

$R_3(\text{DMM}) =$ _____

How do the results compare?

Compare and contrast the two graphs of the series and shunt ohmmeters. Comment accordingly.

Draw a meter scale for your ohmmeter. That is, complete the resistance scale of Fig. 17.5. Is the scale linear or nonlinear?

FIG. 17.5

Part 3   Voltage Divider Ohmmeter

Construct the circuit of Fig. 17.6.

FIG. 17.6

Short the probes and adjust the power supply until the meter movement reads 1 mA.

Separate the probes and read the meter. Record in Table 17.5. Use the DMM to set the potentiometer resistance to the values shown in Table 17.5. Read the meter movement for each resistance setting. Record in Table 17.5.

Plot the data of Table 17.5 using the vertical and horizontal divisions in Graph 17.3.

This graph can now be used as a calibration chart for the voltage divider ohmmeter circuit of Fig. 17.6. Now set the 0–10-kΩ potentiometer (R_x) to three random values, reading the meter each time. Check the results with the DMM.

$R_1 = $ _____ , $R_2 = $ _____ , $R_3 = $ _____

$R_1(\text{DMM}) = $ _____ , $R_2(\text{DMM}) = $ _____ ,

$R_3(\text{DMM}) = $ _____

TABLE 17.5

Resistance Across Probes	$I_{\text{meter movement}}$
0 Ω	1 mA
500 Ω	
1,000 Ω	
2,000 Ω	
3,000 Ω	
4,000 Ω	
5,000 Ω	
8,000 Ω	
10,000 Ω	
∞	0

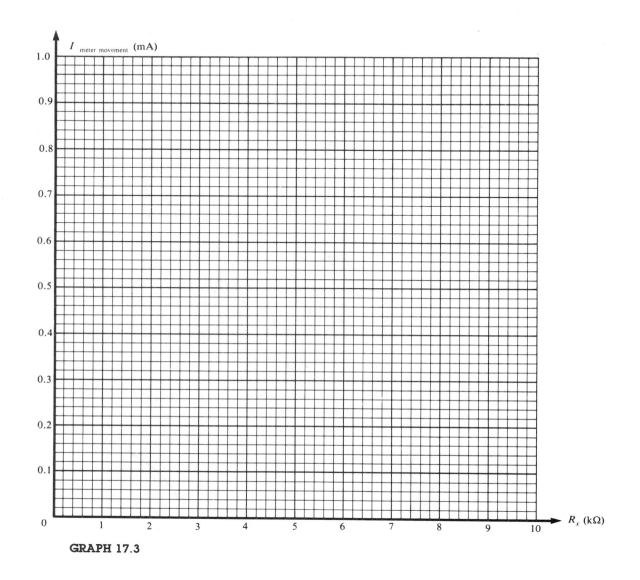

GRAPH 17.3

How do the results compare?

Draw a meter scale for your ohmmeter. That is, complete the resistance scale of Fig. 17.7.

FIG. 17.7

Is the scale linear or nonlinear?

Summarize by comparing and contrasting the three circuits as to their advantages, disadvantages, and characteristics.

Computer Analysis: Series, Parallel, and Series-Parallel Networks

OBJECT

To become familiar with the computer analysis of series, parallel, and series-parallel dc networks.

EQUIPMENT REQUIRED

Computer facilities capable of applying PSPICE or language such as BASIC, PASCAL, FORTRAN, or C.

RÉSUMÉ OF THEORY

The primary purpose of this laboratory exercise is to develop a familiarity with computer techniques as applied to dc network configurations. Although the first part of each section requests the use of PSPICE, if this software package is unavailable the same analysis can be performed in the same section using an appropriate language.

All the networks to be investigated have been constructed and examined in detail in an earlier laboratory session. The computer analysis will be used to verify the results obtained earlier, and of course the earlier results will check the output of the computer runs.

PROCEDURE

Part 1   Series dc Circuit

(a) The series dc network of Fig. 18.1 was examined in Experiment dc 4. From Experiment dc 4, record the measured values of the voltages V_1, V_2, V_3 and the current I.

FIG. 18.1

$V_1 =$ _____, $V_2 =$ _____, $V_3 =$ _____,

$I =$ _____

(b) Redraw the network, and label the various nodes for a PSPICE analysis, using node D as the reference.

(c) Write the input file to determine V_{BD}, V_{CD}, and I.

(d) Perform the PSPICE analysis and record the results.

$V_{BD} =$ _____, $V_{CD} =$ _____, $I =$ _____

(e) How do the results of Part 1(d) compare with the results of Part 1(a), keeping in mind that $V_{BD} = V_2 + V_3$?

(f) Using an appropriate language, write a program to analyze the network of Fig. 18.1 and generate V_1, V_2, V_3, and I.

(g) Run the program of Part 1(f) and record the results.

$V_1 =$ _____, $V_2 =$ _____, $V_3 =$ _____,

$I =$ _____

(h) How do the results of Part 1(g) compare with the results of Part 1(a)?

Part 2 Parallel dc Networks

(a) The parallel dc network of Fig. 18.2 was examined in Experiment dc 6. From Experiment dc 6, record the measured values of I_1, I_2, I_3, and I_S.

FIG. 18.2

$I_1 =$ _____, $I_2 =$ _____, $I_3 =$ _____,

$I_S =$ _____

(b) Redraw the network and label the network as required for a PSPICE analysis.

(c) Write the input file to determine I_1, I_2, I_3, and I_S.

(d) Perform the PSPICE analysis and record the results.

$I_1 =$ ——————— , $I_2 =$ ——————— , $I_3 =$ ——————— ,

$I_S =$ ———————

(e) How do the results of Part 2(d) compare with the results of Part 2(a)?

(f) Using an appropriate language, write a program to analyze the network of Fig. 18.2 and generate I_1, I_2, I_3, and I_S.

(g) Run the program of Part 2(f), and record the results.

$I_1 =$ _____ , $I_2 =$ _____ $I_3 =$ _____ ,

$I_S =$ _____

(h) How do the results of Part 2(g) compare with the results of Part 2(a)?

Part 3   Series-Parallel dc Network

(a) The series-parallel network of Fig. 18.3 was examined in Experiment dc 8. From Experiment dc 8, record the measured values of V_{ad}, V_{bc}, and V_{ab}.

FIG. 18.3

$V_{ad} =$ _____ , $V_{bc} =$ _____ , $V_{ab} =$ _____

(b) Redraw the network, and label the network as required for a PSPICE analysis.

(c) Write the input file to determine V_{ad}, V_{bc}, and V_{ab}.

(d) Perform the PSPICE analysis, and record the results.

$V_{ad} =$ _____, $V_{bc} =$ _____, $V_{ab} =$ _____

(e) How do the results of Part 3(d) compare with the results of Part 3(a)?

(f) Using an appropriate language, write a program to analyze the network of Fig. 18.3 and generate V_{ad}, V_{bc}, and V_{ab}.

(g) Run the program of Part 3(f) and record the results.

$V_{ad} =$ _____, $V_{bc} =$ _____, $V_{ab} =$ _____

(h) How do the results of Part 3(g) compare with the results of Part 3(a)?

Computer Analysis: Superposition, Thevenin's Theorem, Transients

OBJECT

To become familiar with the application of computer techniques to superposition, Thevenin's theorem, and transient analysis.

EQUIPMENT REQUIRED

Computer facilities capable of applying PSPICE or language such as BASIC, PASCAL, FORTRAN, or C.

RÉSUMÉ OF THEORY

This laboratory session with the computer is oriented toward developing innovative methods using PSPICE and providing an increased challenge to those students using a language. The general procedure for the analysis to be performed is outlined in the text and should be referred to before proceeding with this experiment.

Again, all the networks appearing in this experiment have been investigated earlier for comparison purposes.

PROCEDURE

Part 1   Superposition

(a) The network of Fig. 19.1 was analyzed in Experiment dc 9 using superposition. From Experiment dc 9, record the measured values of I_1, I_2, I_3, and V_2.

FIG. 19.1

$I_1 = $ _____ , $I_2 = $ _____ , $I_3 = $ _____ ,

$V_2 = $ _____

(b) Redraw the network, and label the network accordingly for PSPICE analysis.

(c) Write the input file to determine I_1, I_2, I_3, and V_2.

(d) Perform the PSPICE analysis for each source, and record the results below.

E_1: $I_1 =$ _____ , $I_2 =$ _____ , $I_3 =$ _____ ,

$V_2 =$ _____

E_2: $I_1 =$ _____ , $I_2 =$ _____ , $I_3 =$ _____ ,

$V_2 =$ _____

Using the above results, determine the total values of each quantity.

$I_1 =$ _____ , $I_2 =$ _____ , $I_3 =$ _____ ,

$V_2 =$ _____

Perform the PSPICE analysis with both sources present, and record the results below.

$I_1 =$ _____ , $I_2 =$ _____ , $I_3 =$ _____ ,

$V_2 =$ _____

How do the results obtained with both sources present compare with the algebraic sum?

(e) How do the results of the last part of Part 1(d) compare with the results of Part 1(a)?

(f) Using an appropriate language, write a program to determine I_1, I_2, I_3, and V_2 for each source and for both sources present.

(g) Run the program of Part 1(f) and record the results.

E_1: $I_1 =$ _____ , $I_2 =$ _____ , $I_3 =$ _____ ,

$\quad V_2 =$ _____

E_2: $I_1 =$ _____ , $I_2 =$ _____ , $I_3 =$ _____ ,

$\quad V_2 =$ _____

$E_1 + E_2$: $I_1 =$ _____ , $I_2 =$ _____ , $I_3 =$ _____ ,

$\quad V_2 =$ _____

(h) How do the results of Part 1(g) compare with the results of Part 1(a)?

Part 2   Thevenin's Theorem

(a) Thevenin's theorem was applied to the network of Fig. 19.2 in Experiment dc 10. From Experiment dc 10, record the values of E_{Th} and R_{Th}.

FIG. 19.2

$$E_{Th} = \underline{\hspace{3cm}}, \quad R_{Th} = \underline{\hspace{3cm}}$$

(b) Redraw the network of Fig. 19.2, and label the nodes as required by PSPICE to determine the open-circuit voltage $V_{a\text{-}a'}$.

(c) Write the input file to determine $V_{a\text{-}a'}$.

(d) Perform the PSPICE analysis, and record the value of E_{Th}.

$E_{Th} =$ _____

(e) Apply an ideal one-ampere current source to the terminals a-a', and write an input file to determine the Thevenin resistance R_{Th}. Redraw the network, choose the nodes, and write the input file.

(f) Perform the PSPICE analysis, and record the value of R_{Th}.

$R_{Th} =$ _____

(g) Compare the values obtained in Parts 2(d) and 2(f) with those listed in Part 2(a).

(h) Using an appropriate language, write a program to find E_{Th} and R_{Th} for the network of Fig. 19.2.

(i) Run the program of Part 2(h) and record the results.

E_{Th} = _____ , R_{Th} = _____

(j) How do the results of Parts 2(i) and 2(a) compare?

Part 3   Transients

(a) The charging R-C network of Fig. 19.3 was analyzed in Experiment dc 13. Sketch the waveform of V_C and V_R as obtained in Experiment dc 13.

FIG. 19.3

(b) Redraw the network, and label the various nodes for a PSPICE analysis.

(c) Write the input file to obtain a PLOT of V_C and V_R against time.

(d) Perform the PSPICE analysis, and compare the curves obtained with those of Part 3(a).

(e) Using an appropriate language, write a program to obtain a table of V_C vs. time from 0 to 5 τ with data points every 0.1 τ.

(f) Run the program of Part 3(e), and compare the data points with the plot of Part 3(a).

Math Review

ac

The analysis of sinusoidal ac networks will require the extended use of the Pythagorean theorem, trigonometric functions, and vector operations. This review will begin to establish the mathematical foundation necessary to work with each of these functions and operations in a confident, correct, and accurate manner.

It is assumed that a calculator or table is available to determine (with reasonable accuracy) the value of the trigonometric functions (sine, cosine, tangent) and the square and square root of a number.

PYTHAGOREAN THEOREM

The Pythagorean theorem states that the square of the hypotenuse of a right triangle is equal to the sum of the squares of the other sides of the triangle. With the notation in Fig. M.1, the theorem has the following form:

$$\boxed{Z^2 = X^2 + Y^2} \tag{M.1}$$

FIG. M.1

For θ_1,

$$\sin \theta_1 = \frac{\text{opposite}}{\text{hypotenuse}} = \frac{Y}{Z} \tag{M.2}$$

$$\cos \theta_1 = \frac{\text{adjacent}}{\text{hypotenuse}} = \frac{X}{Z} \tag{M.3}$$

$$\tan \theta_1 = \frac{\sin \theta_1}{\cos \theta_1} = \frac{Y/Z}{X/Z} = \frac{Y}{X} = \frac{\text{opposite}}{\text{adjacent}} \tag{M.4}$$

For θ_2,

$$\sin \theta_2 = \frac{X}{Z}$$

$$\cos \theta_2 = \frac{Y}{Z}$$

$$\tan \theta_2 = \frac{X}{Y}$$

To determine θ_1 and θ_2, we use

$$\theta_1 = \sin^{-1} \frac{Y}{Z} = \cos^{-1} \frac{X}{Z} = \tan^{-1} \frac{Y}{X}$$

$$\theta_2 = \sin^{-1} \frac{X}{Z} = \cos^{-1} \frac{Y}{Z} = \tan^{-1} \frac{X}{Y} \tag{M.5}$$

EXAMPLE 1   Find the hypotenuse Z and the angle θ_1 for the right triangle of Fig. M.2.

FIG. M.2

Solution

$$Z = \sqrt{X^2 + Y^2} = \sqrt{(3)^2 + (4)^2} = \sqrt{25} = 5$$

and

$$\theta_1 = \tan^{-1} \frac{4}{3} = \tan^{-1} 1.333 \ldots \cong 53.13°$$

or

$$\theta_1 = \sin^{-1} \frac{4}{5} = \sin^{-1} 0.8 \cong 53.13°$$

or

$$\theta_1 = \cos^{-1} \frac{3}{5} = \cos^{-1} 0.6 \cong 53.13°$$

EXAMPLE 2   Determine Y and θ_1 for the right triangle of Fig. M.3.

FIG. M.3

Solution

$$Z^2 = X^2 + Y^2$$

or

$$Y^2 = Z^2 - X^2$$

and

$$Y = \sqrt{Z^2 - X^2} \qquad \text{(M.6)}$$

Substituting,

$$Y = \sqrt{(12)^2 - (6)^2} = \sqrt{108} \cong 10.39$$

and

$$\theta_1 = \sin^{-1} \frac{X}{Z} = \sin^{-1} \frac{6}{12} = \sin^{-1} 0.5 = 30°$$

PROBLEMS

1. Determine θ_2 for Fig. M.2.

    $\theta_2 = $ _____

2. Determine θ_2 for Fig. M.3.

    $\theta_2 = $ _____

3. **a.** Determine X for the right triangle of Fig. M.4.

    $X = $ _____

FIG. M.4

b. Determine θ_1 and θ_2 for Fig. M.4.

    $\theta_1 = $ _____ , $\theta_2 = $ _____

4. Refer to Fig. M.5.

FIG. M.5

a. Determine Y.

    $Y = $ _____

b. Find X.

    $X = $ _____

c. Determine θ_1.

    $\theta_1 = $ _____

VECTOR REPRESENTATION

There are two common forms for representing vectors with the tail fixed at the origin. The *rectangular* form has the appearance shown in Fig. M.6, where the real and imaginary components define the location of the head of the vector.

FIG. M.6

The imaginary component is defined by the component of the vector that has the *j* associated with it, as appearing in the following representation of the rectangular form:

Rectangular Form:

$$\boxed{\mathbf{Z} = X + jY}$$ (M.7)

In the first and fourth quadrant, X is positive, while in the second and third quadrant, it is negative. Similarly, the imaginary component is positive in the first and second quadrants and negative in the third and fourth.

FIG. M.7

The *polar* form has the appearance shown in Fig. M.7, where the angle θ is *always* measured from the positive real axis. The equation form is as follows:

Polar Form:

$$\boxed{\mathbf{Z} = Z \angle \theta}$$ (M.8)

Through Fig. M.8, it should be fairly obvious that the two forms are related by the following equations.

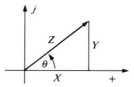

FIG. M.8

Rectangular to Polar:

$$Z = \sqrt{X^2 + Y^2}$$ (M.9)

$$\theta = \tan^{-1} \frac{Y}{X}$$ (M.10)

Polar to Rectangular:

$$X = Z \cos \theta$$ (M.11)

$$Y = Z \sin \theta$$ (M.12)

EXAMPLE 3   Determine the rectangular and polar forms of the vector of Fig. M.9.

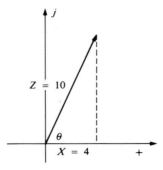

FIG. M.9

Solution   Rectangular form:

$$Y = \sqrt{Z^2 - X^2} = \sqrt{(10)^2 - (4)^2} = \sqrt{84} = 9.165$$

and

$$\mathbf{Z} = 4 + j9.165$$

Polar form:

$$\theta = \tan^{-1} \frac{Y}{X} = \tan^{-1} \frac{9.165}{4} = \tan^{-1} 2.291 = 66.42°$$

and

$$\mathbf{Z} = 10 \angle 66.42°$$

EXAMPLE 4   Determine the polar form of the vector of Fig. M.10.

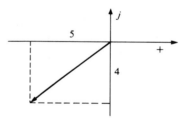

FIG. M.10

Solution   In conversions of this type, it is usually best to isolate a right triangle, as shown in Fig. M.11, and determine θ' and Z.

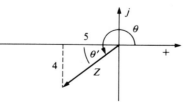

FIG. M.11

The angle θ' is not the angle θ of the polar form but it can be determined from the right triangle and then used to calculate θ.

In this case,

$$Z = \sqrt{X^2 + Y^2} = \sqrt{(5)^2 + (4)^2} = \sqrt{41} \cong 6.4$$

and

$$\theta' = \tan^{-1} \frac{4}{5} = \tan^{-1} 0.8 = 38.66°$$

Using Fig. M.11, we find

$$\theta = 180° + \theta' = 180° + 38.66° = 218.66°$$

and

$$\mathbf{Z} = 6.4 \angle 218.66°$$

PROBLEMS

5. Determine the polar form of each of the following vectors:

   a. $\mathbf{Z} = 10 + j10 = $ _____

   b. $\mathbf{Z} = -2 + j8 = $ _____

   c. $\mathbf{Z} = -50 - j200 = $ _____

   d. $\mathbf{Z} = +0.4 - j0.8 = $ _____

6. Determine the rectangular form of each of the following vectors:

   a. $\mathbf{Z} = 6 \angle 37.5° = $ _____

   b. $\mathbf{Z} = 2 \times 10^{-3} \angle 100° = $ _____

   c. $\mathbf{Z} = 52 \angle -120° = $ _____

   d. $\mathbf{Z} = 1.8 \angle -30° = $ _____

ADDITION AND SUBTRACTION OF VECTORS

The sum or difference of vectors is normally found in the rectangular form. The operation can be performed in the polar form only if the vectors have the same angle or are out of phase by 180°.

Addition

Rectangular Form:

$$(X_1 + jY_1) + (X_2 + jY_2) = (X_1 + X_2) + j(Y_1 + Y_2) \tag{M.13}$$

Polar Form:

$$Z_1 \angle \theta + Z_2 \angle \theta = (Z_1 + Z_2) \angle \theta \tag{M.14}$$

Subtraction

Rectangular Form:

$$(X_1 + jY_1) - (X_2 + jY_2) = (X_1 - X_2) + j(Y_1 - Y_2) \qquad \textbf{(M.15)}$$

Polar Form:

$$Z_1 \angle\theta - Z_2 \angle\theta = (Z_1 - Z_2) \angle\theta \qquad \textbf{(M.16)}$$

EXAMPLE 5   Determine the sum of the following vectors:

$$\mathbf{X}_1 = 4 + j4 \qquad \mathbf{X}_2 = 6 \angle120°$$

Solution   Using Fig. M.12 to convert $\mathbf{X}_2$ to rectangular form,

$$Y = 6 \sin 60° = 6(0.866) = 5.196$$
$$X = 6 \cos 60° = 6(0.5) = 3$$

and

$$\mathbf{Z} = 6 \angle120° = -3 + j5.196$$

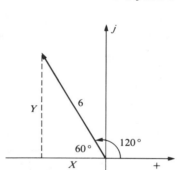

FIG. M.12

Note in the above conversion that the proper sign for X and Y is simply determined by the location of the vector.

$$\mathbf{X}_{\text{sum}} = \mathbf{X}_1 + \mathbf{X}_2 = (4 + j4) + (-3 + j5.196)$$

Note the use of parentheses around each vector to identify clearly each vector and the sign of each term.

$$\mathbf{X}_{\text{sum}} = (4 - 3) + j(4 + 5.196) = 1 + j9.196$$

EXAMPLE 6   Determine the difference between the following vectors:

$$\mathbf{X}_1 = -5 - j4 \qquad \mathbf{X}_2 = 3 - j8$$

Solution

$$\mathbf{X}_{\text{difference}} = \mathbf{X}_1 - \mathbf{X}_2 = (-5 - j4) - (3 - j8) = (-5 - 3) + j(-4 + 8)$$

Carefully review the sign of each term of the above equation and note how the parentheses were an aid in generating the proper signs for the following solution:

$$\mathbf{X}_{\text{difference}} = -8 + j4$$

PROBLEMS

7. Perform the following vector operations:

   a. $(10 + j30) + (-4 - j8) =$ _____

   b. $(10 \angle 45°) - (7.07 - j7.07) =$ _____

   c. $(50 \angle 60°) + (0.8 \angle 60°) =$ _____

   d. $(16 \angle -20°) + (14 \angle 160°) =$ _____

   e. $(5000 + j1000) - (2500 \angle -60°) =$ _____

   f. $(5 \angle 0°) + (20 \angle -90°) - (6 \angle 180°) =$ _____

MULTIPLICATION AND DIVISION OF VECTORS

Multiplication

Multiplication of vectors is routinely accomplished in either form, although the polar form is more direct.

Polar Form:

$$(Z_1 \angle \theta_1)(Z_2 \angle \theta_2) = Z_1 Z_2 \angle \theta_1 + \theta_2 \qquad \textbf{(M.17)}$$

Rectangular Form:

There is a lengthy equation for multiplication in the rectangular form but it seems more logical to simply state that the operation is carried out by simply multiplying the real and imaginary part of one vector by both components of the other vector.

EXAMPLE 7 Determine the product of the following vectors:

$$X_1 = 10 \angle 50° \qquad X_2 = 0.4 \angle -60°$$

Solution:

$$X_{product} = X_1 \cdot X_2 = (10 \angle 50°)(0.4 \angle -60°) = (10)(0.4) \angle 50° + (-60°) = 4 \angle -10°$$

EXAMPLE 8 Calculate the product of the following vectors:

$$X_1 = 4 + j6 \qquad X_2 = 0.2 - j0.6$$

Solution:

$$\begin{array}{r} 4 + j6 \\ \times\ 0.2 - j0.6 \\ \hline -j2.4 - j^2 3.6 \\ +0.8 + j1.2 \\ \hline \end{array}$$ (multiplying both terms of X_1 by $-j0.6$)
(multiplying both terms of X_1 by 0.2)

Since $j^2 = -1$, the sum of the terms is the following:

$$0.8 + j(1.2 - 2.4) - (-1)3.6 = 0.8 + 3.6 - j1.2$$

and

$$\mathbf{X}_{product} = 4.4 - j1.2$$

Division

Although division of vectors is performed more directly in the polar form, the operation can also be performed in the rectangular form:

Polar Form:

$$\boxed{\frac{Z_1 \angle \theta_1}{Z_2 \angle \theta_2} = \frac{Z_1}{Z_2} \underline{/\theta_1 - \theta_2}} \qquad \text{(M.18)}$$

Rectangular Form:

Division in this form is best performed by first using the following useful equation derived by multiplying the vector in the denominator by its complex conjugate (change the sign of the imaginary part):

$$\frac{1}{X + jY} = \frac{1}{X + jY}\frac{(X - jY)}{(X - jY)} = \frac{X - jY}{X^2 + Y^2} = \frac{X}{X^2 + Y^2} - j\frac{Y}{X^2 + Y^2} \quad \text{(M.19)}$$

Note that the result has the complex conjugate of the original vector in the numerator and a divisor equal to the sum of the squares of the two components of the rectangular form.

Division then becomes a multiplication process, as shown below:

$$\frac{X_1 + jY_1}{X_2 + jY_2} = (X_1 + jY_1)\left(\frac{1}{X_2 + jY_2}\right) = (X_1 + jY_1)\left(\frac{X_2}{X_2^2 + Y_2^2} - j\frac{Y_2}{X_2^2 + Y_2^2}\right)$$

EXAMPLE 9 Perform the following operations:

(a) $\dfrac{\mathbf{X}_1}{\mathbf{X}_2}$ if $\mathbf{X}_1 = 80 \angle 10°$ and $\mathbf{X}_2 = 40 \angle -20°$

(b) $\dfrac{4 - j9}{3 + j4}$

Solutions

(a) $\mathbf{X}_{division} = \dfrac{\mathbf{X}_1}{\mathbf{X}_2} = \dfrac{80 \angle 10°}{40 \angle -20°} = \dfrac{80}{40} \underline{/10 - (-20°)} = 2 \underline{/10 + 20°}$

$= 2 \angle 30°$

Note the need to bracket the angle from the denominator to be sure the proper sign is obtained.

(b) $\dfrac{4 - j9}{3 + j4} = (4 - j9)\left(\dfrac{3}{3^2 + 4^2} - j\dfrac{4}{3^2 + 4^2}\right) = (4 - j9)\left(\dfrac{3}{25} - j\dfrac{4}{25}\right)$

$\qquad = \dfrac{12}{25} - j\dfrac{16}{25} - j\dfrac{27}{25} + j^2\dfrac{36}{25}$

$\qquad = \left(\dfrac{12}{25} - \dfrac{36}{25}\right) + j\left(\dfrac{-16}{25} - \dfrac{27}{25}\right)$

$\qquad = \dfrac{-24}{25} - j\dfrac{43}{25}$

eliminating the need to convert to the polar form before the division could take place.

PROBLEMS

8. Perform the following operations:

 a. $(50 \angle 60°)(4 \angle 80°) = $ _____

 b. $(0.8 \angle -10°)(3 \angle 30°)(-4 \angle 20°) = $ _____

 c. $(5 + j8)(4 - j2) = $ _____

 d. $(48 \angle 20°)(10 - j1) = $ _____

 e. $(0.9 - j0.5)(40 + j10) = $ _____

 f. $(4000 \angle 0°)(2000 \angle -20°)(1000 + j1000) = $ _____

9. Perform the following operations:

 a. $\dfrac{1000 \angle 60°}{20 \angle -50°} = $ _____

 b. $\dfrac{0.008 \angle 50°}{5 \times 10^{-6} \angle 100°} = $ _____

 c. $\dfrac{6 + j6}{2 + j2} = $ _____

 d. $\dfrac{10 \angle 30°}{4 + j10} = $ _____

 e. $\dfrac{20,000 + j10,500}{5000 \angle -50°} = $ _____

 f. $\dfrac{(50 \angle 20°)(20 \angle -30°)}{(5 + j5)(4 \angle -10°)} = $ _____

The Oscilloscope

OBJECT

To introduce the basic components and use of an oscilloscope and audio oscillator (or function generator).

EQUIPMENT REQUIRED

Instruments

1 – DMM (or VOM)

1 – Oscilloscope

1 – Audio oscillator or function generator

Miscellaneous

2 – D batteries with holders

EQUIPMENT ISSUED

TABLE 1.1

Item	Manufacturer and Model No.	Laboratory Serial No.
DMM (or VOM)		
Oscilloscope		
Audio oscillator or function generator		

RÉSUMÉ OF THEORY

Oscilloscope

The cathode-ray oscilloscope is an instrument that displays the variations of a voltage with time on a *cathode-ray tube* (CRT) such as that appearing in Fig. 1.1. The CRT is an evacuated glass envelope, shaped as shown in Fig. 1.1, with the following essential components:

FIG. 1.1   **Cathode-ray tube**

1. An electron gun that generates rapidly moving electrons, shapes them into a pencil-like beam, and then directs them along the axis of the tube
2. Deflection plates for producing a vertical or horizontal deflection of the beam from its original path
3. A fluorescent screen that emits light at the point where the beam strikes

The *electron gun* is made up of the following components:

1. A heater and cathode that serve as the source of the electrons
2. A control electrode that serves to vary the strength of the beam current
3. A focusing electrode that focuses the beam to a sharp point on the screen
4. Accelerating and preaccelerating electrodes that give the electrons the high velocity required to reach the screen and cause secondary emission

The deflection of the beam may be accomplished either electrostatically or electromagnetically. Most oscilloscopes use electrostatic deflection (Fig. 1.1). To deflect the beam, a potential is applied across plates D_1 and D_2, or D_3 and D_4. The horizontal plates D_1 and D_2 deflect the beam vertically, whereas the vertical plates D_3 and D_4 deflect the beam horizontally. The deflection of the beam is directly proportional to the impressed voltage.

The CRT is, in itself, not a very sensitive device. A 5-in. CRT, under ordinary conditions, will give about 1 in. of deflection for a difference of potential of 100 V. The signals encountered most frequently are usually well under 100 V, so it becomes necessary to amplify the signals before a usable deflection will be obtained. Two deflection amplifiers are required, one for each set of plates.

For the oscilloscope to indicate the variations of an electrical quantity in the vertical direction as a function of time, there must be a voltage impressed on the horizontal deflection plates that varies linearly with time. This voltage is shown graphically in Fig. 1.2. An oscillator that can generate such a voltage is called, for obvious reasons, a *sawtooth* oscillator. Frequently, the term *sweep* oscillator is used. Figure 1.3 provides a block diagram of a general-purpose oscilloscope.

FIG. 1.2

FIG. 1.3

All of the controls for the proper operation of an oscilloscope are mounted on the front panel of the instrument. Figure 1.4 indicates the approximate locations of the controls found on most general-purpose oscilloscopes. The locations of the controls shown vary according to manufacturer. Table 1.2 describes the function of each.

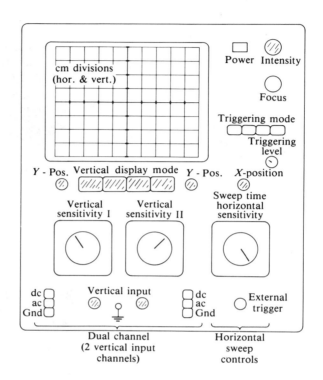

FIG. 1.4

TABLE 1.2

Control	Function
Power	Turns on the main power.
Intensity	Controls the intensity of the pattern on the screen.
Focus	Focuses the electron beam so that the pattern will be clearly defined.
Triggering mode	Determines the type of triggering for the horizontal sweeping pattern.
Triggering level	Determines the level at which triggering should occur.
Vertical display mode	Determines whether one or two signals will be displayed at the same time and which technique will be used to display the signals.
Y-position	Controls the vertical location of the pattern.
X-position	Controls the horizontal location of the pattern.
Vertical sensitivity	Determines the volts/cm for the vertical axis of display.
Sweep time horizontal sensitivity	Determines the time/cm for the horizontal axis of display.
dc/ac/Gnd switch	Determines whether dc levels will be displayed on the screen and permits determining the Gnd (zero-volt input level) of the display.

(a) Voltage Measurements

dc Levels:

To use the scope to measure dc levels, first place the dc/ac/Gnd switch in the Gnd position to establish the ground (zero-volt) level on the screen.

Then switch the dc/ac/Gnd switch to the dc position to measure the dc level. In the ac mode, a capacitor blocks the dc from the screen.

Then place the scope leads across the unknown dc level and use the following equation to determine the dc level:

$$dc\ Level\,(V) = deflection\,(cm) \times vertical\ sensitivity\,(V/cm)$$ **(1.1)**

ac Levels:

After re-establishing the ground level, place the dc/ac/Gnd switch in the ac mode and connect the scope leads across the unknown voltage. The peak-to-peak voltage can then be determined from

$$V_{p\text{-}p}(V) = deflection\ peak\ to\ peak\,(cm) \times vertical\ sensitivity\,(V/cm)$$ **(1.2)**

(b) Frequency Measurements The oscilloscope can be used to set the frequency of an audio oscillator or function generator using the *horizontal sensitivity* in the following manner.

Determine the period of the desired waveform and then calculate the number of divisions required to display the waveform on the horizontal axis using the provided μs/cm, ms/cm, or s/cm on the horizontal sensitivity control. Then adjust the audio oscillator or function generator to provide the proper horizontal deflection for the desired frequency.

Of course, the reverse of the above procedure will determine the frequency of an unknown signal.

Audio Oscillator and Function Generator

The *audio oscillator* is designed to provide a sinusoidal waveform in the frequency range *audible* by the human ear. A *function generator* typically expands on the capabilities of the audio oscillator by providing a square wave and triangular waveform with an increased frequency range. Either instrument is suitable for this experiment since we will be dealing only with sinusoidal waveforms in the audio range.

Most oscillators and generators require that the magnitude of the output signal be set by an oscilloscope or DMM (or VOM). That is, the amplitude dial of the oscillator is not graduated and the peak or peak-to-peak value is set by connecting the output of the oscillator to a scope or meter and adjusting the amplitude dial until the desired voltage is obtained.

PROCEDURE

Part 1 Introduction

(a) Your instructor will introduce the basic operation of the oscilloscope and audio oscillator or function generator.

(b) Turn on the oscilloscope and establish a horizontal line centered on the face of the screen. There are no connections to the vertical input sections of the scope for this part.

(c) Adjust the controls listed in Table 1.3 and comment on the effects.

TABLE 1.3

Control	Observed Effect
Focus	
Intensity	
Y-position	
X-position	

Part 2   dc Voltage Measurements

(a) Set the dc/ac/Gnd switch to the Gnd position and adjust the Y-position control until the *zero-volt* reference is a line centered vertically on the screen.

(b) Once the zero-volt level is established, move the dc/ac/Gnd switch to the dc position and set the vertical sensitivity to 1 V/cm and connect one channel of the scope across the 1.5-V battery as shown in Fig. 1.5.

FIG. 1.5

Record the vertical shift below.

Vertical shift = _____ cm

Determine the voltage by multiplying by the vertical sensitivity. That is,

dc Voltage = (vertical shift)(vertical sensitivity)

= (_____)(_____)

= _____ V

Change the sensitivity to 0.5 V/cm and note the effect on the vertical shift. Recalculate the dc voltage with this new shift.

dc Voltage = (vertical shift)(vertical sensitivity)

= (_____)(_____)

= _____ V

How do the two measurements compare?

Which is more accurate? Why?

(c) Disconnect the 1.5-V battery and re-establish the zero-volt reference line. Then connect the vertical input section of the scope as shown in Fig. 1.6 with the vertical sensitivity set at 1 V/cm.

FIG. 1.6

What was the effect on the magnitude and direction of the shift?

Can a scope determine whether a particular voltage is positive or negative? How?

Calculate the measured voltage as follows:

dc Voltage = (vertical shift)(vertical sensitivity)

= (_____)(_____)

= _____ V

Measure the voltage with the DMM (or VOM) and compare results.

dc Voltage [DMM (or VOM)] = _____ V

Part 3   Sinusoidal Waveforms—Magnitude

In this part of the experiment we will learn how to set the magnitude of a sinusoidal signal using an oscilloscope or DMM (or VOM). The frequency will remain fixed at 500 Hz.

Oscilloscope:

(a) Connect the output of the oscillator or generator directly to one channel of the scope as shown in Fig. 1.7.

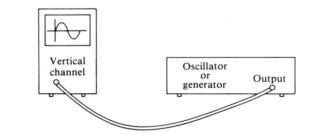

FIG. 1.7

(b) Set the output frequency of the oscillator or generator to 500 Hz using the dial and appropriate multiplier. Turn the amplitude knob all the way to the left for minimum output.

(c) Set the vertical sensitivity of the scope to 1 V/cm and the horizontal sensitivity to 0.5 ms/cm and turn on both the scope and the oscillator or generator.

(d) Set the dc/ac/Gnd switch to the Gnd position to establish the zero-volt reference level (also the vertical center of a sinusoidal waveform) and then return the switch to the ac position.

(e) Now adjust the amplitude control of the oscillator or generator until the signal has a 4-V peak-to-peak swing. The resulting waveform has the following mathematical formulation:

$$v = V_m \sin 2\pi ft = 2 \sin 2\pi 500t$$

(f) Switch to the dc position and comment below on any change in the waveform.

(g) Make the necessary adjustments to display the following waveforms on the screen. Sketch both patterns in the space provided, showing the number of divisions (in centimeters) for the vertical and horizontal distances, and the vertical and horizontal sensitivities.

   1. $v = 0.2 \sin 2\pi 500t$

Vertical Sensitivity = _____

Horizontal Sensitivity = _____

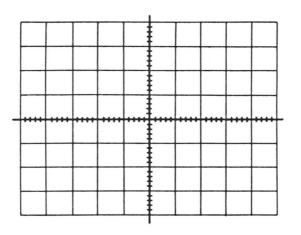

   2. $v = 8 \sin 2\pi 500t$

Vertical Sensitivity = _____

Horizontal Sensitivity = _____

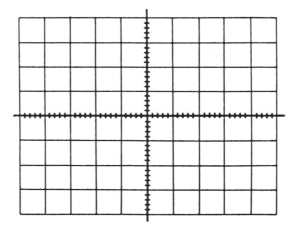

DMM (or VOM):

   (h) The sinusoidal signal $2 \sin 2\pi 500t$ has an effective value determined by

$$V_{\text{eff}} = 0.707 V_m = 0.707(2) = 1.414 \text{ V}$$

Connect the DMM (or VOM) directly across the oscillator in the ac rms mode and adjust the oscillator output until $V_{eff} = 1.414$ V. Then connect the output of the oscillator directly to the scope and note the total peak-to-peak swing.

Is the waveform the same as that obtained in part 3(e)?

(i) Use the DMM (or VOM) to set the following sinusoidal output from the oscillator:

$$v = 5 \sin 2\pi 500t$$

$V_{eff} =$ _____ V

Set V_{eff} with the DMM (or VOM) by adjusting the output of the oscillator, and place the signal on the screen.

Calculate the peak-to-peak voltage as follows:

$$V_{p\text{-}p} = (\text{vertical distance peak to peak})(\text{vertical sensitivity})$$

$= ($ _____ $)($ _____ $)$

$=$ _____ V

How does the above compare with the desired 10-V peak-to-peak voltage?

Part 4   Sinusoidal Waveforms – Frequency

This section will demonstrate how the oscilloscope can be used to set the frequency output of an oscillator or generator. In other words, the scope can be used to make fine adjustments on the frequency set by the dials of the oscillator or generator.

For a signal such as $2 \sin 2\pi 500t$, the frequency is 500 Hz and the period is $1/500 = 2$ ms. With a horizontal sensitivity of 0.5 ms/cm, the waveform should appear in exactly four horizontal divisions. If it does not, the fine-adjust control on the frequency of the oscillator or generator can be adjusted until it is exactly 4 cm. The scope has then set the output frequency of the oscillator.

Make the necessary adjustments to place the following waveforms on the scope. Sketch the waveforms in the space provided, indicating the number of vertical and horizontal deflections and the sensitivity of each.

1. $v = 0.4 \sin 62{,}832t$

$f =$ _____ Hz, $T =$ _____ s

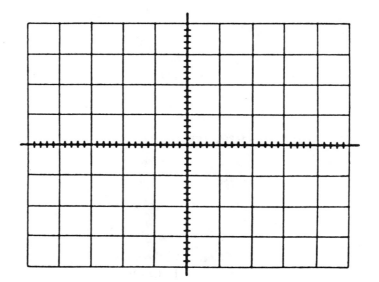

vertical deflection = _____ cm,

vertical sensitivity = _____

horizontal deflection = _____ cm,

horizontal sensitivity = _____

2. $v = 5 \sin 377t$

$f =$ _____ Hz, $T =$ _____ s

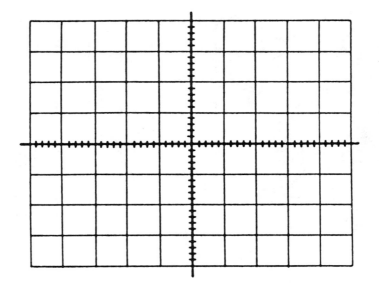

vertical deflection = _____ cm,

vertical sensitivity = _____

horizontal deflection = _____ cm,

horizontal sensitivity = _____

Part 5   Sinusoidal Waveforms on a dc Level

(a) Set the oscillator or generator to an output of $1 \sin 2\pi 500t$ using a vertical sensitivity of 1 V/cm on the scope with a horizontal sensitivity of 0.5 ms/cm.

(b) Measure the dc voltage of one of the D cells and insert in Fig. 1.8.

$E =$ _____ V

(c) Construct the network of Fig. 1.8.

FIG. 1.8

(d) The input signal now has a dc level equal to the dc voltage of the D cell.
Set the dc/ac/Gnd switch to the Gnd position and adjust the zero line to the center of the screen.

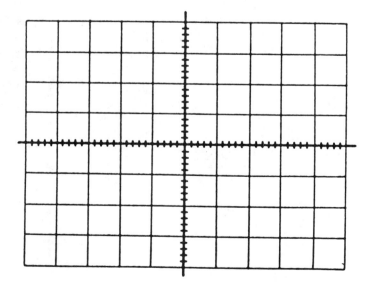

(e) Switch the ac mode and sketch the response, clearly indicating the zero-level line and the vertical deflection in centimetres.

(f) Switch to the dc mode and sketch the response, clearly indicating the zero-volt reference level and the vertical deflections in centimetres.

(g) How are the waveforms of parts 5(e) and 5(f) different? Why?

R-L-C Components

OBJECT

To measure the impedance of a resistor, inductor, and capacitor at a fixed frequency.

EQUIPMENT REQUIRED

Resistors

1 — 100-Ω, 1.2-kΩ, 3.3-kΩ

Inductors

2 — 10-mH

Capacitors

1 — 0.5-μF, 1-μF

Instruments

1 — DMM (or VOM)

1 — Oscilloscope

1 — Audio oscillator (or signal generator)

EQUIPMENT ISSUED

TABLE 2.1

Item	Manufacturer and Model No.	Laboratory Serial No.
DMM (or VOM)	BK Persision 2830	03095
Oscilloscope	Leader 1021	03933
Audio oscillator (or signal generator)	Circuitmate (F62)	03834

TABLE 2.2

Resistors	
Nominal Value	Measured Value
100 Ω	100.5
1.2 kΩ	1.24 K
3.3 kΩ	3.34 K Ω

RÉSUMÉ OF THEORY

For impedances in series, the total impedance is the sum of the individual impedances:

$$\boxed{\mathbf{Z}_T = \mathbf{Z}_1 + \mathbf{Z}_2 + \mathbf{Z}_3 + \cdots + \mathbf{Z}_N}$$

(2.1)

For resistors in series,

$$\boxed{R_T = R_1 + R_2 + R_3 + \cdots + R_N}$$

(2.2)

which is independent of frequency.

For inductors in series,

$$X_{L_T} = X_1 + X_2 + X_3 + \cdots + X_N$$
$$= 2\pi f L_1 + 2\pi f L_2 + 2\pi f L_3 + \cdots + 2\pi f L_N$$
$$= 2\pi f L_T$$

and

$$\boxed{L_T = L_1 + L_2 + L_3 + \cdots + L_N}$$

(2.3)

Note that the individual and total inductive reactances are directly proportional to frequency.

For capacitors in series,

$$X_{C_T} = X_1 + X_2 + X_3 + \cdots + X_N$$

$$= \frac{1}{2\pi f C_1} + \frac{1}{2\pi f C_2} + \frac{1}{2\pi f C_3} + \cdots + \frac{1}{2\pi f C_N}$$

$$= \frac{1}{2\pi f C_T}$$

where

$$\boxed{\frac{1}{C_T} = \frac{1}{C_1} + \frac{1}{C_2} + \frac{1}{C_3} + \cdots + \frac{1}{C_N}}$$
(2.4)

Note that the individual and total capacitive reactances are inversely proportional to frequency.

Consider the special case of only two capacitors in series:

$$X_{C_T} = \frac{1}{2\pi f C_T}$$

where

$$C_T = \frac{C_1 C_2}{C_1 + C_2}$$

For resistors in parallel,

$$\boxed{\frac{1}{R_T} = \frac{1}{R_1} + \frac{1}{R_2} + \frac{1}{R_3} + \cdots + \frac{1}{R_N}}$$
(2.5)

which is independent of frequency.

For inductors in parallel,

$$\frac{1}{X_{L_T}} = \frac{1}{X_1} + \frac{1}{X_2} + \frac{1}{X_3} + \cdots + \frac{1}{X_N}$$

$$= \frac{1}{2\pi f L_1} + \frac{1}{2\pi f L_2} + \frac{1}{2\pi f L_3} + \cdots + \frac{1}{2\pi f L_N}$$

$$= \frac{1}{2\pi f L_T}$$

where

$$\boxed{\frac{1}{L_T} = \frac{1}{L_1} + \frac{1}{L_2} + \frac{1}{L_3} + \cdots + \frac{1}{L_N}}$$
(2.6)

Consider the special case of only two inductors in parallel:

$$X_{L_T} = \frac{X_1 X_2}{X_1 + X_2} = 2\pi f L_T$$

where

$$X_1 = 2\pi f L_1 \qquad X_2 = 2\pi f L_2 \qquad L_T = \frac{L_1 L_2}{L_1 + L_2}$$

For capacitors in parallel,

$$\frac{1}{X_{C_T}} = \frac{1}{X_1} + \frac{1}{X_2} + \frac{1}{X_3} + \cdots + \frac{1}{X_N}$$

$$= 2\pi f C_1 + 2\pi f C_2 + 2\pi f C_3 + \cdots + 2\pi f C_N$$

$$= 2\pi f C_T$$

where

$$\boxed{C_T = C_1 + C_2 + C_3 + \cdots + C_N} \qquad\qquad (2.7)$$

Consider the special case of only two capacitors in parallel:

$$X_{C_T} = \frac{X_1 X_2}{X_1 + X_2} = \frac{1}{2\pi f C_T}$$

where

$$C_T = C_1 + C_2$$

PROCEDURE

Part 1   Resistance

(a) Construct the circuit of Fig. 2.1. Insert the actual values of the resistors as determined by the ohmmeter section of your multimeter.

FIG. 2.1

Caution: Always ensure that the ground of the oscilloscope is connected to the ground of the oscillator. Otherwise a hazardous situation may result.

(b) Set the voltage across R to 4 V (*p-p*) using the oscilloscope. Measure the rms voltage across the sensing resistor (100 Ω) with the DMM (or VOM).

$V_{R_S} = $ _____.04 V_____

Calculate the peak value of V_{R_S}

$$(\sqrt{2})(.04v) = .05898$$

$V_p = $ _____.05898_____

Calculate the peak-to-peak value of V_{R_S}

$$2(\sqrt{2})(.041v) = .1159$$

$V_{p-p} = $ _____.1159_____

(c) Calculate the peak-to-peak value of the current I from

$$I_{p-p} = \frac{V_{R_{S(p-p)}}}{R_S}$$

$$\frac{.1159v}{100\Omega} = 1.15mA$$

$I_{p-p} = $ _____1.15 mA_____

(d) Determine the resistance of the resistor R from

$V_R = 1.376(\sqrt{2})$

$$R = \frac{4\text{ V }(p\text{-}p)}{I_{p-p}}$$

$$\frac{4v}{1.15mA} = 3.478K\Omega$$

$R = $ _____3.478K_____

(e) Compare the value obtained in part 1(d) with the measured value inserted in Fig. 2.1.

*) Calculated value in (d) is 138K higher due to the voltage Drop At V_S

Table 2.1 ⇒ 3.44K Ω

(f) Connect a 1.2-kΩ and a 3.3-kΩ resistor in series as shown in Fig. 2.2

$R_{S\text{ measured}} =$ 100.5 Ω
$R_{1\text{ measured}} =$ 1.21 kΩ
$R_{2\text{ measured}} =$ 3.34 kΩ

$+ V_{R_S} -$

R_S

Audio oscillator or signal generator $f = 1$ kHz

$+$ V $-$

100 Ω I

R_1 1.2 kΩ

$V_R = 4$ V $(p\text{-}p)$

R_2 3.3 kΩ Oscilloscope

FIG. 2.2

Calculate the total resistance of the series combination of the 1.2-kΩ and 3.3-kΩ resistors using the measured values.

$$\frac{4.65 - 4.7}{4.65}$$

$R_T =$ 4.65 k

(g) Set the voltage across the series combination to 4 V ($p\text{-}p$) and then measure the voltage V_{R_S} with the DMM or VOM.

$V_{R_S} =$ .03

(h) Calculate the peak-to-peak value of V_{R_S}.

$$(.03)(\sqrt{2})(2) = .08485$$

$V_{R_S(p\text{-}p)} =$ .08485

(i) Determine the peak-to-peak value of the current I from

$$I_{p\text{-}p} = \frac{V_{R_S(p\text{-}p)}}{R_S} = \frac{.08485}{100.5\,\Omega} = .844\,mA$$

$I_{p\text{-}p} =$ .844 mA

(j) Calculate the total resistance R_T of the series combination from

$$R_T = \frac{V_{R_{T(p\text{-}p)}}}{I_{p\text{-}p}} = \frac{4\text{ V}}{I_{p\text{-}p}} \qquad \frac{4V}{.844\,mA} =$$

$R_T = $ _4.739K_

(k) Compare the results of part 1(j) with the calculated value of part 1(f).

★ The

Part 2 Capacitive Reactance

(a) Construct the circuit of Fig. 2.3. Insert the measured value of R_S.

$R_{S\text{ measured}} = $ _100.5_

.805 uF

$V_c = 4$ V (p-p)

FIG. 2.3

(b) Set the voltage V_C to 4 V (*p-p*) and measure the rms value of the voltage V_{R_S} with the DMM (or VOM).

$V_{R_S} = $ _.687_

(c) Using the technique described in part 1, calculate the peak-to-peak value of the current *I*.

$$2(\sqrt{2})(.687 v) = 1.94 v_{p-p}$$

$V_{RSp-p} = 1.94 V.$

$$\frac{1.94 v}{100.5 \Omega} = 19.33 mn$$

$I_{p-p} = $ _19.33 MA_

(d) Calculate the reactance X_C from the peak-to-peak values of V_C and *I*.

$$X_C = \frac{V_{C(p-p)}}{I_{p-p}}$$

$\dfrac{3.81}{19.33 MA}$

$V_c = 1.35$

$(1.35)(\sqrt{2})(2)$

$X_C = 197.54$

X_C(measured) = $\underline{197.54\ \Omega}$

(e) Using the nameplate value of the capacitance, calculate the reactance of the capacitor at $f = 1$ kHz and compare with the results of part 2(d).

$$X_c = \frac{1}{(2\pi)(1k)(1\mu F)} \quad 159.15 \qquad C = 1\mu F \qquad \frac{1}{2\pi f C}$$
$$f = 1k H$$

X_C(calculated) = $\underline{159.15}$

(f) Using the measured value of X_C in part 2(d), determine the capacitance level if $f = 1$ kHz and compare with the nameplate value of the capacitance.

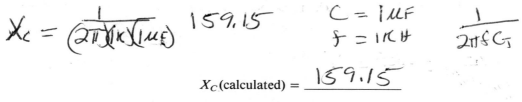

$$X_C = 197.54 \quad \frac{1}{(2\pi)(1k)C} = \frac{1}{(2\pi)(1k)(197.54)}$$

$$C = 805\,\mu F \qquad C \qquad = 805\,\mu F$$

(g) Connect a 0.5-μF in parallel with the capacitor of Fig. 2.3 and set V_C to 4 V (*p-p*) again. Measure the rms value of V_{R_S} using a DMM (or VOM).

V_{R_S} = $\underline{1.087}$

Calculate the peak-to-peak value of V_{R_S}.

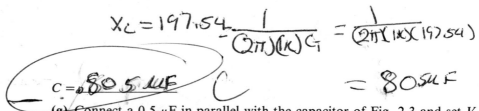

$$(1.087)(2)(\sqrt{2})$$

$V_{R_{S(p-p)}}$ = $\underline{3.07}$

Determine the peak-to-peak value of I from

$$I_{p-p} = \frac{V_{R_{S(p-p)}}}{R_S} = \frac{3.07 V}{100.6\Omega} = 30.5 mA$$

I_{p-p} = $\underline{30.5\,mA}$

(h) Calculate the total reactance of the parallel capacitors from

$$X_{C_T} = \frac{V_{C(p\text{-}p)}}{I_{p\text{-}p}} = \frac{4\text{ V}}{I_{p\text{-}p}} \qquad = \frac{4v}{30.5mA} = 131.15\,\Omega$$

X_{C_T}(measured) = $131.15\,\Omega$

(i) Using the total capacitance as determined from the nameplate values, calculate the total reactance of the parallel capacitor at a frequency of $f = 1$ kHz.

$$C_T = .5\mu F + 1\mu F = 1.5\mu F \qquad .333\ \mu F, \quad \frac{1}{(2\pi)(1K)(1.5\mu F)}$$

X_{C_T}(calculated) = $106.18\,\Omega$

Compare the result with the measured value of part 2(h).

$\boxed{1.405\mu F}$

(j) Using the measured value of X_{C_T} in part 2(h), calculate the equivalence total capacitance at $f = 1$ kHz and compare with the nameplate level.

$$C_T = \frac{1}{(2\pi \times 1K)(131.15)}$$

$C_T = 1.21\mu F$

Part 3   Inductive Reactance

(a) Construct the network of Fig. 2.4. Insert the measured values for R_S and R_l.

$R_{S\text{ measured}} = 100.00$

$R_{l\text{ measured}} = 11.8\,\Omega$

$V_L = 2$ V $(p\text{-}p)$

FIG. 2.4

(b) Set the voltage V_L to 2 V (p-p) and measure the rms value of the voltage V_{R_S} with a DMM (or VOM).

$V_{R_S} = \underline{\hspace{0.5em}1.12v\hspace{1em}}$

(c) Calculate $V_{R_{S(p-p)}}$

$$2(\sqrt{2})(1.12v) \doteq 3.17v$$

$V_{R_{S(p-p)}} = \underline{\hspace{0.5em}3.17V\hspace{1em}}$

(d) Calculate I_{p-p}

$$\frac{V_{R_S p-p}}{R_s} = \frac{3.17v}{100 s} = 31.5 ma$$

$I_{p-p} = \underline{\hspace{0.5em}31.5 ma\hspace{1em}}$

(e) Ignoring the effects of R_l, calculate the reactance X_L from

$$X_L = \frac{V_{L(p-p)}}{I_{p-p}} = \frac{2 \text{ V}}{I_{p-p}}$$

$$X_L = \frac{2v}{31.5ma}$$

X_L (measured) $= \underline{\hspace{0.5em}63.5\,\Omega\hspace{1em}}$

(f) Using the nameplate value of inductance L, calculate the inductive reactance at $f = 1$ kHz.

$$X_L = 2\pi f \, 10mH = (2\pi)(1k)(10mH)$$

X_L (calculated) $= \underline{\hspace{0.5em}62.83\hspace{1em}}$

(g) Compare the results of part 3(f) with the measured level of part 3(e).

(h) Calculate the inductance level necessary to establish an inductive reactance of the magnitude measured in part 3(e).

$$63.5 = (2\pi)(1K)(L_T) = \frac{63.5\,\Omega}{(2\pi)(1K)} = L_T$$
$$= 10.1\,m\text{-}H$$

$L = \underline{10.1\ m\text{-}H}$

(i) Compare the results of part 3(h) with the nameplate inductance level.

error of only $\boxed{.1\ m\text{-}H}$

(j) Connect a second 10-mH coil in series with the coil of Fig. 2.4 and establish 2 V (*p-p*) across the two series coils. Then measure the rms value of V_{R_S} and calculate $V_{R_S(p\text{-}p)}$.

$R_2 = 12\,\Omega$

$$V_{R_{S(rms)}} = \underline{\quad .52\,\Omega \quad}$$
$$V_{R_{S(p\text{-}p)}} = \underline{\quad 1.47V \quad}$$

$2(2)(.52v)$

(k) Calculate $I_{p\text{-}p}$.

$$\frac{1.47}{100.5} = 14.63$$

$I_{p\text{-}p} = \underline{14.63\ mA}$

(l) Determine X_{L_T} from

$$X_{L_T} = \frac{V_{L(p\text{-}p)}}{I_{p\text{-}p}} = \frac{2\ V}{I_{p\text{-}p}}$$

$\dfrac{2v}{14.63\ mA}$

$X_{L_T}(\text{measured}) = \underline{136.76}$

(m) Using the nameplate values of the inductances, calculate the total reactance at $f = 1$ kHz and compare with the results of part 3(l).

$$L_T = \frac{1}{10\,m\text{-}H} + \frac{1}{10\,m\text{-}H} = 5\,m\text{-}H$$

$$X_{L_T}(\text{calculated}) = \underline{ 5 \; m-H }$$

(n) Using the measured value of X_{L_T} in part 3(l), calculate the required inductance level at $f = 1$ kHz and compare with the total nameplate inductance level.

$$\frac{1}{X_{LT}} = \frac{1}{2(\pi)(1k)(.15 m-H)}$$

$$136.66 =$$

Frequency Response of R and L Components

OBJECT

To note the effect of frequency on the basic R and L components.

EQUIPMENT REQUIRED

Resistors

1—47-Ω, 100-Ω, 1-kΩ

Inductors

2—10-mH

Instruments

1—DMM

1—Oscilloscope

1—Audio oscillator (or function generator)

EQUIPMENT ISSUED

TABLE 3.1

Item	Manufacturer and Model No.	Laboratory Serial No.
DMM	BK pessision	03105
Oscilloscope	Leader 1021	03933
Audio oscillator (or function generator)	Circuitmate CFG2	03743

TABLE 3.2

Resistors	
Nominal Value	**Measured Value**
47 Ω	48.2
100 Ω	100.6
1 kΩ	.999K

RÉSUMÉ OF THEORY

The resistance of a carbon resistor is unaffected by frequency, except for extremely high frequencies. This rule is also true for the total resistance of resistors in series or parallel.

The reactance of an inductor is linearly dependent on the frequency applied. That is, if we double the frequency, we double the reactance, as determined by $X_L = 2\pi f L$. For very low frequencies, the reactance is correspondingly very small, while for increasing frequencies, the reactance will increase to a very large value. For dc conditions, we find that $X_L = 2\pi(0)L$ is zero ohms, corresponding with the short-circuit representation we used in our analysis of dc circuits. For very high frequencies, X_L is so high that we can often use an open-circuit approximation.

PROCEDURE

Part 1   Resistors

Construct the circuit of Fig. 3.1. Insert the measured value of R.

In this part of the experiment, the voltage across the resistor will be held constant while only the frequency is varied. If the resistance is frequency independent, the current through the circuit should not change as a function of frequency. Therefore, by keeping the voltage V_{ab} constant and changing the frequency while monitoring the current I, we can verify if, indeed, resistance is frequency independent.

Set the voltage V_{ab} to 4 V (p-p) (measure this voltage with the oscilloscope). Then set the frequencies to those shown in Table 3.3, each time monitoring V_{ab} [always 4 V (p-p)] and I. Since we are using the DMM for current measurement, this value is given in rms. Record the values in Table 3.3.

FIG. 3.1

TABLE 3.3

Frequency	$V_{ab(p-p)}$	$V_{ab(rms)} = \dfrac{0.707}{2} V_{ab(p-p)}$	I_{rms}	$R = \dfrac{V_{ab(rms)}}{I_{rms}}$
100 Hz	4 V	1.414 V	1.404 mA	1000 Ω
200 Hz	4 V	1.414 V	1.401 mA	1009 Ω
500 Hz	4 V	1.414 V	1.4 mA	1010 Ω
1000 Hz	4 V	1.414 V	1.398 mA	1018 Ω
2000 Hz	4 V	1.414 V	1.382 mA	1023 Ω

Did the resistance *R* as calculated in the last column change with frequency?

Suspected an increase in resistance in DMM as frequency increases

Note Pg 521
resistive Breakedown

Part 2   Inductors

(a) Construct the circuit of Fig. 3.2. (Measure the dc resistance of the coil with the DMM and record that value as R_l on the diagram.)

In this part, the resistor of part 1 is replaced by the inductor. Here again, the voltage across the inductor will be kept constant while we vary the frequency of that volt-

FIG. 3.2

$R_1 = 12.8 \, \Omega$

age and monitor the current in the circuit. The 47-Ω resistor ensures that there is some resistance in the circuit for all frequency levels to limit the current.

Set the voltage V_{ab} to 4 V (p-p) (use the oscilloscope to measure V_{ab}) and the oscillator or generator to the various frequencies shown in Table 3.4, each time making sure that $V_{ab} = 4$ V (p-p) and that you measure the current I (remember I will be in rms). Record in Table 3.4.

(handwritten: 1.2 mH   2 mH)

TABLE 3.4

Frequency	$V_{ab(p\text{-}p)}$	$V_{ab(rms)}$	I_{rms}	$X_L\text{(measured)} = \dfrac{V_{ab(rms)}}{I_{rms}}$	$X_L\text{(calculated)} = 2\pi fL$
1 kHz	4 V	1.414 V	5.1 mA	277.25	62.83
3 kHz *(3030)*	4 V	1.414 V	14.1 mA	100.3	190.4
5 kHz *(5000)*	4 V	1.414 V	22.1 mA	63.98	314.
7 kHz	4 V	1.414 V	30.3 mA	46.67	440
10 kHz	4 V	1.414 V	40.6 mA	34.82	628

(b) Calculate the reactance X_L (magnitude only) at each frequency and insert the values in Table 3.4 under the heading "X_L(measured)". We are assuming that $X_L \gg R_l$ for the frequencies of interest, so that

$$Z = \frac{V_{ab(rms)}}{I_{rms}} = \sqrt{R_l^2 + X_L^2} \cong \sqrt{X_L^2} = X_L$$

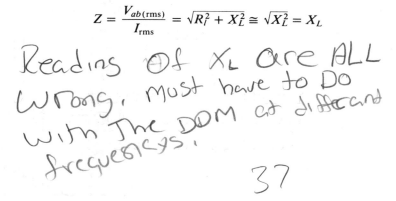

(handwritten: Reading of X_L are ALL wrong, must have to Do with The DMM at different frequencys.   37)

(c) Calculate the reactance at each frequency of Table 3.4 using the nameplate value of inductance (10 mH), and complete the table.

(d) Plot the measured and calculated values of X_L versus frequency on Graph 3.1. Label each curve and plot the points accurately.

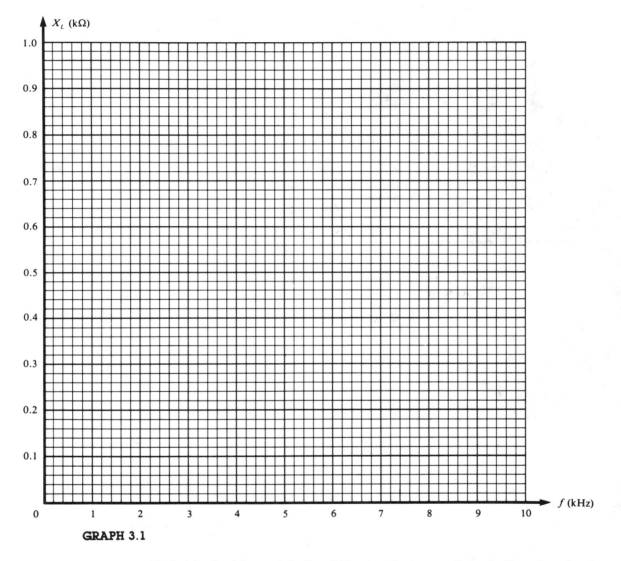

GRAPH 3.1

(e) Are both plots straight lines? Do they both pass through $X_L = 0$ at $f = 0$ Hz as they should?

(f) Determine the inductance at 1500 Hz using the results of part 2(d). That is, determine X_L from the graph and calculate L from $L = X_L/2\pi f$. Compare with the nameplate value.

$$L_{\text{calculated}} = \underline{\hspace{4cm}} \text{ , } L_{\text{nameplate}} = \underline{\hspace{4cm}}$$

Part 3   The Sensing Resistor

Whenever possible, the use of milliammeters is avoided in practice because the network has to be disturbed to insert the meter. In place of the meter is placed a sensing resistor that through Ohm's Law permits the calculation of the circuit current. In Fig. 3.3 a 100-Ω sensing resistor has been added in place of the DMM of Fig. 3.2. In addition, two coils are in parallel to change the net inductance level to $L_T = 10 \text{ mH}/2 = 5 \text{ mH}$. The effect of R_l is ignored in this part.

$R_1 = 11.91$

$R_2 = 12.8$

$R_{S \text{ measured}} = \underline{100.6}$

FIG. 3.3

(a) Construct the network of Fig. 3.3. With the generator set at 1 kHz, adjust the output of the supply until a 4-V peak-to-peak signal is obtained across the parallel coils. As noted in Part 2, the rms value of the voltage across the coils will be 1.414 V.

(b) For each frequency appearing in Table 3.5, set the voltage across the coils to 4 V (*p-p*) and measure the rms value of the voltage across the resistor R_S. Insert the values of $V_{R_{S(\text{rms})}}$ in Table 3.5.

TABLE 3.5

$V_{ab} = 1.414$

$L = 5 mH$

Frequency	$V_{ab(\text{rms})}$	$V_{R_{S(\text{rms})}}$	I_{rms}	$X_L \text{(measured)} = \dfrac{V_{ab(\text{rms})}}{I_{\text{rms}}}$	$X_L \text{(calculated)} = 2\pi fL$
1 kHz	1.414 V	4.25	42.25 mA	33.47 Ω	31.42
3 kHz	1.414 V	1.47	14.6 mA	96.78 Ω	94.25
5 kHz	1.414 V	.88	8.75 mA	161.6	157.1
7 kHz	1.414 V	.64	6.36 mA	222.33	219.91
10 kHz	1.414 V	.44	4.37 mA	323.57	314.16

330

(c) For each frequency of Table 3.5, calculate the rms value of the current from

$$I_{rms} = \frac{V_{R_{S(rms)}}}{R_S(\text{measured})}$$

and insert in Table 3.5.

(d) Calculate X_L from the measured values at each frequency and insert in the X_L(measured) column of Table 3.5.

(e) Calculate X_L using the total nameplate inductance level of 5 mH and insert in the X_L(calculated) column.

(f) How do the results of the X_L(measured) column compare with the X_L(calculated) column?

Considerable close

(g) Plot X_L (measured) versus frequency on Graph 3.1. Is the graph a straight line? What was the impact of a reduced inductance level on the location of the graph?

(h) Determine X_L at a frequency of 6 kHz from the graph.

$X_L =$ _____

Calculate the equivalent inductance level at this frequency.

$L =$ _____

Frequency Response of Capacitive Components

OBJECT

*To note the effect of frequency on the reactance
of a capacitor.*

EQUIPMENT REQUIRED

Resistors

$1-100\text{-}\Omega$

Capacitors

$1-0.1\text{-}\mu\text{F},\ 0.5\text{-}\mu\text{F}$

Instruments

$1-$ DMM
$1-$ Oscilloscope
$1-$ Audio oscillator (or function generator)

EQUIPMENT ISSUED

TABLE 4.1

Item	Manufacturer and Model No.	Laboratory Serial No.
DMM		
Oscilloscope		
Audio oscillator (or function generator)		

TABLE 4.2

Resistors	
Nominal Value	Measured Value
100 Ω	100.6

RÉSUMÉ OF THEORY

The effect of frequency on the impedance of a resistor and inductor was examined in Experiment 3. We will now investigate the change in the reactance of a capacitor with an increase in frequency. At low frequencies the reactance of a coil is quite low, whereas for a capacitor the reactance is quite high at low frequencies, often permitting the use of an open-circuit equivalent. At higher frequencies the reactance of a coil increases rapidly in a linear fashion, the reactance of a capacitor decreases in a nonlinear manner. In fact, it drops off more rapidly than the reactance of a coil increases. At very high frequencies the capacitor can be approximated by a short-circuit equivalency.

PROCEDURE

Part 1 X_C vs f

(a) Construct the circuit of Fig. 4.1.

The 100-Ω sensing resistor will be used to "sense" the current level in the network. With the generator set at 200 Hz, adjust the output of the supply until a 4-V

FIG. 4.1

peak-to-peak signal is obtained across the capacitor. The resulting rms voltage across the capacitor is then 1.414 V rms.

(b) For each frequency appearing in Table 4.3, set the voltage across the capacitor to 4 V (*p-p*) and measure the rms value of the voltage across the resistor R_S. Insert the values of $V_{R_{S(rms)}}$ in Table 4.3.

TABLE 4.3

Frequency	$V_{C(rms)}$	$V_{R_{S(rms)}}$	$I_{(rms)}$	X_C(measured) $= \dfrac{V_{C(rms)}}{I_{(rms)}}$	X_C(calculated) $= \dfrac{1}{2\pi fC}$
200 Hz	1.414 V	.017	.169 mA	8366.86	7957.75
500 Hz	1.414 V	.043	.427 mA	3311.48	3183.1
800 Hz	1.414 V	.067	.666 mA	2123.12	1989.44
1000 Hz	1.414 V	.086	.855 mA	1653.8	1591.55
2000 Hz	1.414 V	.167	1.66 mA	851.81	795.77

(c) For each frequency in Table 4.3, calculate the rms value of the current from

$$I_{rms} = \frac{V_{R_{S(rms)}}}{R_S(\text{measured})}$$

and insert in Table 4.3.

(d) Calculate X_C from the measured values at each frequency and insert in the X_C(measured) column of Table 4.3.

(e) Calculate X_C using the nameplate capacitance level of 0.1 μF at each frequency and insert in the X_C(calculated) column.

(f) How do the results in the X_C(measured) column compare with those in the X_C(calculated) column?

Both reactance's Reduce porportionally with with an increase in frequency The values are slitly different due to not being able to get an exact meas. value for cap.

(g) Plot X_C(measured) versus frequency on Graph 4.1. Is the graph linear or nonlinear? Where does the greatest change in capacitive reactance occur (in what frequency range)?

(h) Determine X_C at a frequency of 650 Hz from the curve just plotted.

$X_C =$ _____

At this frequency, determine the equivalent capacitance level using $X_C = 1/2\pi fC$

$C =$ _____

How does this level compare with the nameplate level of 0.1 μF?

Part 2 Parallel Capacitors (Increase in Capacitance)

(a) Place a 0.5-μF capacitor in parallel with the 0.1-μF capacitor of Fig. 4.1 to obtain a nameplate capacitance level of 0.6 μF ($C_T = C_1 + C_2$).

(b) Maintaining 4 V (p-p) across the parallel capacitors, measure and insert the rms value of V_{R_S} for each frequency in Table 4.4 using the technique described in part 1.

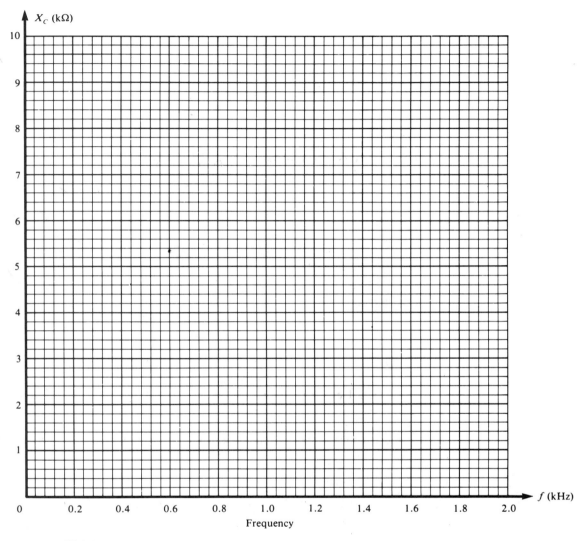

GRAPH 4.1

TABLE 4.4

Frequency	$V_{C(rms)}$	$V_{R_{S(rms)}}$	$I_{(rms)}$	$X_C\text{(measured)} = \dfrac{V_{C(rms)}}{I_{(rms)}}$	$X_C\text{(calculated)} = \dfrac{1}{2\pi fC}$
200 Hz	1.414 V	.116	1.15mA	1229.57	1326.29
500 Hz	1.414 V	.279	2.72mA	510.47 Ω	530.52
800 Hz	1.414 V	.432	4.29	329.6 Ω	331.57
1000 Hz	1.414 V	.543	5.4mA	261.85 Ω	265.26
2000 Hz	1.414 V	1.01	10.04mA	141.26 Ω	132.6

(c) For each frequency in Table 4.4, calculate the rms value of the current from

$$I_{rms} = \frac{V_{R_{S(rms)}}}{R_S(\text{measured})}$$

and insert in Table 4.4.

(d) Calculate X_C from the measured values at each frequency and insert in the X_C(measured) column of Table 4.4.

(e) Calculate X_C using the nameplate capacitance level of 0.6 μF at each frequency and insert in Table 4.4 in the X_C(calculated) column.

(f) How do the results in the X_C(measured) column compare with those in the X_C(calculated) column?

(g) Plot X_C(measured) versus frequency on Graph 4.1. How did the change (increase) in capacitance level affect the location and characteristics of the curve?

(h) Determine X_C at a frequency of 1.5 kHz from the curve just plotted.

$X_C =$ _____

At this frequency, determine the equivalent capacitance level using $X_C = 1/(2\pi fC)$.

$C =$ _____

How does this capacitance level compare with the nameplate level of 0.6 μF?

Part 3 Series Capacitors (Reduced Capacitance Level)

(a) Place the 0.5-μF capacitor in series with the 0.1-μF capacitor of Fig. 4.1 to obtain a nameplate capacitance level of

$$C_T = \frac{(0.5\ \mu\text{F})(0.1\ \mu\text{F})}{0.5\ \mu\text{F} + 0.1\ \mu\text{F}} = 0.083\ \mu\text{F}$$

(b) Maintaining 4 V (*p-p*) across the series combination of capacitors, measure and insert the rms value of V_{R_S} for each frequency in Table 4.5 using the technique described in part 1.

(c) For each frequency in Table 4.5, calculate the rms value of the current from

$$I_{rms} = \frac{V_{R_{S(rms)}}}{R_S(\text{measured})}$$

and insert in Table 4.5.

TABLE 4.5

1083 μF

Frequency	$V_{C(rms)}$	$V_{R_{S(rms)}}$	$I_{(rms)}$	$X_C(\text{measured}) = \dfrac{V_{C(rms)}}{I_{(rms)}}$	$X_C(\text{calculated}) = \dfrac{1}{2\pi fC}$
200 Hz	1.414	.017	.169 ma	8366.86	9587.65
500 Hz	1.414	.042	.417 ma	3390.89	3835.0
800 Hz	1.414	.066	.656 ma	2155.49	2396.91
1000 Hz	1.414	.085	.845 ma	1673.37	1917.53
2000 Hz	1.414	.172	1.72 ma	822.09	958.76

(d) Calculate X_C from the measured values at each frequency and insert in the X_C(measured) column of Table 4.5.

(e) Calculate X_C using the nameplate capacitance level of 0.083 μF at each frequency and insert in Table 4.5 in the X_C(calculated) column.

(f) How do the results in the X_C(measured) column compare with those in the X_C(calculated) column?

(g) Plot X_C(measured) versus frequency in Graph 4.1. How did the change (decrease) in capacitance level affect the location and characteristics of the curve?

(h) Determine X_C at a frequency of 1.8 kHz from the curve just plotted.

$X_C = $ _____

At this frequency, determine the equivalent capacitance level using $X_C = 1/(2\pi fC)$.

$C = $ _____

How does this capacitance level compare with the nameplate level of 0.083 μF?

The Oscilloscope and Phase Measurements

OBJECT

To use the oscilloscope to determine the phase angle between two sinusoidal waveforms.

EQUIPMENT REQUIRED

Resistors

1—1-kΩ

1—0–10-kΩ potentiometer

Capacitors

1—0.5-μF

Instruments

1—DMM

1—Oscilloscope

1—Audio oscillator or function generator

EQUIPMENT ISSUED

TABLE 7.1

Item	Manufacturer and Model No.	Laboratory Serial No.
DMM		
Oscilloscope		
Audio oscillator or function generator		

TABLE 7.2

Resistors	
Nominal Value	**Measured Value**
1 kΩ	

PHASE MEASUREMENTS

The phase angle between two signals of the same frequency can be determined using the oscilloscope. There are two methods available.

    1.   Dual-trace comparison with the calibrated time base
    2.   Lissajous pattern

Dual-Trace Method of Phase Measurement

The dual-trace method of phase measurement, aside from providing a high degree of accuracy, can compare two signals of different amplitudes and, in fact, different wave shapes. The method can be applied directly to oscilloscopes equipped with two vertical channels or to a conventional single-trace oscilloscope with an external electronic switch, as shown in Fig. 7.1. The electronic switch will switch between inputs at a very high speed, so both patterns will appear on the screen.

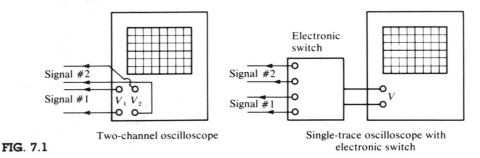

Two-channel oscilloscope   Single-trace oscilloscope with
electronic switch

FIG. 7.1

Regardless of which oscilloscope is available, the procedure essentially consists of displaying both traces on the screen simultaneously, and measuring the distance (in scale divisions) between two identical points on the two traces (Fig. 7.2).

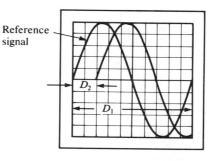

Reference signal

D_2

D_1

FIG. 7.2

One signal will be chosen as a reference, that is, zero-phase angle. In the comparison, therefore, we can assume that the signal being compared is leading ($+\theta$) if it is to the left of the reference and lagging ($-\theta$) if it is to the right of the reference. To use the dual-trace phase measurement method, therefore, proceed as follows:

1. Connect the two signals to the two vertical channels, making sure to observe proper grounding.
2. Select the mode of operation—"Alternate" or "Chop." For frequencies less than 50 kHz, use "Chop." For frequencies greater than 50 kHz, use "Alternate."
3. Once the traces are on the screen, use the Gnd switch to set both patterns in the vertical center of the screen.
4. Measure the number of horizontal divisions (in centimetres) required for one full cycle of either waveform (D_1 in Fig. 7.2).
5. Determine the scale factor SF from

$$SF = \frac{360°}{D_1} \qquad \text{(degrees/division)} \qquad (7.1)$$

6. Determine the number of horizontal divisions (in centimetres) between corresponding positive slopes of the waveforms, that is, the phase shift between the waveforms (D_2 in Fig. 7.2).
7. Calculate the phase angle from

$$\theta = (SF)(D_2) \qquad (7.2)$$

or

$$\theta = \left(\frac{D_2}{D_1}\right) 360° \qquad (7.3)$$

Lissajous-Pattern Phase Measurement

The Lissajous-pattern method is also called the X-Y phase measurement. To use this method, proceed as follows:

1. Connect the two signals to the vertical and the horizontal inputs, respectively.
2. A pattern known as a *Lissajous* will appear on the screen. The type of pattern will indicate certain common phase relationships, and in fact the patterns can be used to calculate the phase angle in general.

The patterns shown in Fig. 7.3 indicate the phase relationship appearing with each figure.

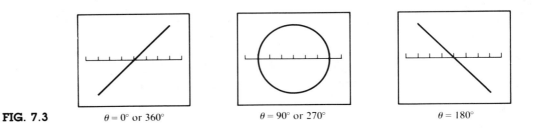

FIG. 7.3 $\theta = 0°$ or $360°$ $\theta = 90°$ or $270°$ $\theta = 180°$

The patterns shown in Fig. 7.4 can be used to calculate the phase angle (θ) as indicated below the figures.

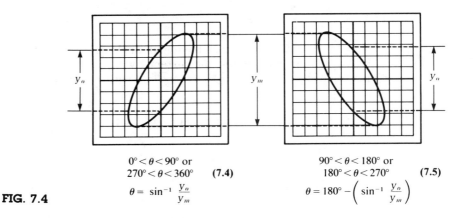

$$0° < \theta < 90° \text{ or}$$
$$270° < \theta < 360° \quad (7.4)$$

$$\theta = \sin^{-1} \frac{y_o}{y_m}$$

$$90° < \theta < 180° \text{ or}$$
$$180° < \theta < 270° \quad (7.5)$$

$$\theta = 180° - \left(\sin^{-1} \frac{y_o}{y_m} \right)$$

FIG. 7.4

EXAMPLE Assume that the patterns in Figs. 7.5 and 7.6 appear on an oscilloscope screen. Calculate the phase angle θ in each case.

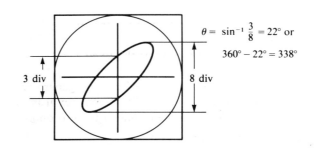

$$\theta = \sin^{-1} \frac{3}{8} = 22° \text{ or}$$

$$360° - 22° = 338°$$

3 div   8 div

FIG. 7.5

$$\theta = 180° - \left(\sin^{-1} \frac{5}{10} \right) = 150° \text{ or}$$
$$360° - 150° = 210°$$

5 div   10 div

FIG. 7.6

PROCEDURE

Part 1

The phase relationship between $\mathbf{E}$ and $\mathbf{V}_R$ of Fig. 7.7 will be determined in this part using the dual-trace capability of the oscilloscope.

$$\mathbf{E} = E \angle 0°$$ $$\mathbf{V}_R = V_R \angle \theta_1$$

FIG. 7.7

Since $\mathbf{E}$ is defined as having an angle of zero degrees, it will appear as shown in Fig. 7.8. In an *R-C* circuit, the current $\mathbf{I}$ will lead the applied voltage as shown in the phasor diagram. The voltage $\mathbf{V}_R$ is in phase with $\mathbf{I}$, and the voltage $\mathbf{V}_C$ will lag the voltage $\mathbf{E}$.

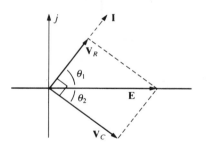

FIG. 7.8

Note that $\theta_1 + \theta_2 = 90°$ and that the vector sum of $\mathbf{V}_R$ and $\mathbf{V}_C$ equals the applied voltage $\mathbf{E}$.

Construct the network of Fig. 7.9, designed to place the voltage $\mathbf{V}_R$ across channel #1 of the scope and $\mathbf{E}$ across channel #2. Note that the resistance R is composed of a series combination of the fixed 1-kΩ resistor and 10-kΩ potentiometer. That is, $R = 1 \text{ k}\Omega + 10 \text{ k}\Omega$.

The oscillator or function generator is set to 200 Hz with an amplitude of 8 V (*p-p*). Set the amplitude and frequency using channel #2 of the oscilloscope.

Set the potentiometer to zero ohms and calculate the reactance of the capacitor at a frequency of 200 Hz and insert in each row of Table 7.3.

FIG. 7.9

TABLE 7.3

Potentiometer Setting	R	X_C	$E_{\text{rms}} = 0.707V_p$	$V_{R(\text{rms})}$		θ_1	
				Calculated	Measured	Calculated	Measured
0 Ω	1000 Ω		2.828 V				
2000 Ω	3000 Ω		2.828 V				
4000 Ω	5000 Ω		2.828 V				

     (a) Using $R = 1000$ Ω, the calculated value of X_C, and $\mathbf{E} = 2.828 \angle 0°$, calculate the rms value of $\mathbf{V}_R$ and the angle θ_1 associated with the resistance as defined in Fig. 7.8 and insert in Table 7.3.

     In general,

$$|V_R| = \frac{RE}{\sqrt{R^2 + X_C^2}} \qquad (7.6)$$

with

$$\theta_1 = \tan^{-1}\frac{X_C}{R} \qquad (7.7)$$

$R = 1000$-Ω Calculations:

Repeat the above for $R = 3000\ \Omega$ and $5000\ \Omega$ and insert the calculated results in Table 7.3.

$R = 3000$-Ω Calculations:

$R = 5000$-Ω Calculations:

(b)

$R = 1000$-Ω Measurements:

Measure $V_{R(p\text{-}p)}$ from the scope display and calculate $V_{R(\text{rms})}$ and insert in Table 7.3 for $R = 1000\ \Omega$.

$$V_{R(p\text{-}p)} = \underline{\hspace{3cm}}\ ,\ V_{R(\text{rms})} = \underline{\hspace{3cm}}$$

Determine the number of horizontal divisions for one full cycle of **E** or **V$_R$**.

Number of horizontal divisions (in centimetres) for one full cycle =

$D_1 = \underline{\hspace{3cm}}$

Determine the number of horizontal divisions representing the phase shift between waveforms.

Number of horizontal divisions (in centimetres) between waveforms =

$D_2 = \underline{\hspace{3cm}}$

Determine the phase shift in degrees from

$$\theta_1 = \left(\frac{D_2}{D_1}\right) 360° \qquad \textbf{(7.8)}$$

$\theta_1 = ($ _____ $) \ 360° = $ _____ $°$

Insert the measured value of θ_1 in Table 7.3.

       Repeat the above for $R = 3000 \ \Omega$ and $5000 \ \Omega$ and insert the results in Table 7.3.

$R = 3000$-Ω Measurements:

$V_{R(p\text{-}p)} = $ _____ , $V_{R(\text{rms})} = $ _____

$D_1 = $ _____ cm, $D_2 = $ _____ cm

$\theta_1 = $ _____ $°$

$R = 5000$-Ω Measurements:

$V_{R(p\text{-}p)} = $ _____ , $V_{R(\text{rms})} = $ _____

$D_1 = $ _____ cm, $D_2 = $ _____ cm

$\theta_1 = $ _____ $°$

    (c) Calculate the percent difference between the calculated and measured values of θ_1 from Table 7.3 using the following equation, and insert in Table 7.4.

$$\% \text{ Difference } (\theta_1) = \frac{|\text{calculated} - \text{measured}|}{\text{calculated}} \times 100\% \qquad \textbf{(7.9)}$$

TABLE 7.4

R	% Difference in θ_1
1000 Ω	
3000 Ω	
5000 Ω	

(d) The vector $\mathbf{E} = E \angle \theta = 2.828$ V $\angle 0°$ has been placed on each phasor diagram of Graph 7.1. Note that the voltage has been scaled to match the 1 V/cm scale of the horizontal and vertical axes. Using the measured values of $V_{R(\text{rms})}$ and θ_1, insert the phasor $\mathbf{V}_R$ for each value of R. Clearly indicate the angle θ_1 and the magnitude of $V_{R(\text{rms})}$.

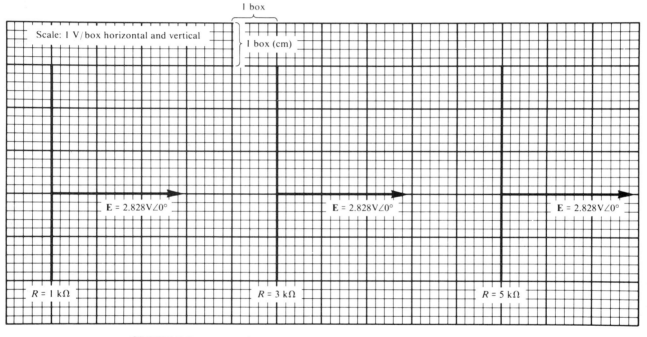

GRAPH 7.1

Part 2

The phase relationship between $\mathbf{E}$ and $\mathbf{V}_C$ of the same network as Fig. 7.7 will now be determined by interchanging the resistor R and capacitor C, as shown in Fig. 7.10.

FIG. 7.10

The elements must exchange positions to avoid a "shorting out" of the resistor R of Fig. 7.7 if the oscilloscope were simply placed across the capacitor of Fig. 7.7. The grounds of the scope and supply would establish a zero-volt drop across R and possibly high currents in the remaining network since X_C is the only impedance to limit the current level.

The connections are now made as shown in Fig. 7.11. Note that **E** remains on channel #2 and $\mathbf{V}_C$ is placed on channel #1. The phase angle θ_2 between **E** and $\mathbf{V}_C$ can therefore be determined from the display.

FIG. 7.11

Insert the reactance of the capacitor on each line of Table 7.5 and set the supply to an 8 V (p-p) signal at 200 Hz.

TABLE 7.5

Potentiometer Setting		R	X_C	$E_{rms} = 0.707\,V_p$	$V_{C(rms)}$		θ_2	
					Calculated	**Measured**	**Calculated**	**Measured**
0 Ω		1000 Ω		2.828 V				
2000 Ω		3000 Ω		2.828 V				
4000 Ω		5000 Ω		2.828 V				

(a) Using $R = 1000\ \Omega$, the calculated value of X_C, and $E = 2.828$ V $\angle 0°$, calculate the rms value of V_C and the angle θ_2 associated with the reactance as defined by Fig. 7.8 and insert in Table 7.5.

In general,

$$|V_C| = \frac{X_C E}{\sqrt{R^2 + X_C^2}} \tag{7.10}$$

with

$$\theta_2 = -90° + \tan^{-1}\frac{X_C}{R} \tag{7.11}$$

$R = 1000$-Ω Calculations:

Repeat the above for $R = 3000\ \Omega$ and $5000\ \Omega$ and insert the calculated results in Table 7.5.

$R = 3000$-Ω Calculations:

$R = 5000$-Ω Calculations:

(b)

$R = 1000$-Ω Measurements:

Measure $V_{C(p-p)}$ from the scope display and calculate $V_{C(rms)}$ and insert in Table 7.5 for $R = 1000\ \Omega$.

$V_{C(p\text{-}p)} =$ _____ , $V_{C(\text{rms})} =$ _____

Determine the phase shift in the same manner as described for part 1.

$D_1 =$ _____ cm, $D_2 =$ _____ cm

$\theta_2 =$ _____ °

Repeat the above for $R = 3000\ \Omega$ and $5000\ \Omega$ and insert the results in Table 7.5.

$R = 3000\text{-}\Omega$ Measurements:

$V_{C(p\text{-}p)} =$ _____ , $V_{C(\text{rms})} =$ _____

$D_1 =$ _____ cm, $D_2 =$ _____ cm

$\theta_2 =$ _____ °

$R = 5000\text{-}\Omega$ Measurements:

$V_{C(p\text{-}p)} =$ _____ , $V_{C(\text{rms})} =$ _____

$D_1 =$ _____ cm, $D_2 =$ _____ cm

$\theta_2 =$ _____ °

(c) Complete the phasor diagrams of Graph 7.1 with the insertion of the phasor V_C for each resistance level.

(d) It was noted early in part 1 that $\theta_1 + \theta_2 = 90°$. For each resistance level, add the two and determine the percent difference using the equation

$$\% \text{ Difference} = \frac{|90° - \theta_T|}{90°} \times 100\% \qquad (7.12)$$

Record the values in Table 7.6.

TABLE 7.6

R	θ_1 (part 1)	θ_2 (part 2)	$\theta_T = \theta_1 + \theta_2$	% **Difference 90° vs.** θ_T
1000 Ω				
3000 Ω				
5000 Ω				

Part 3

In this part, the phase angle will be determined from the Lissajous pattern.

Construct the network of Fig. 7.12, which is exactly the same as Fig. 7.9 except now the applied voltage **E** is placed on channel #1 with V_R applied as a *horizontal input*.

FIG. 7.12

The resulting Lissajous patterns will determine the phase angle between **E** and V_R for various values of R.

Set the value of R as indicated in Table 7.7 and measure the y-intercept (y_o) and the y-maximum (y_m) from the Lissajous pattern. Record in Table 7.7.

TABLE 7.7

Potentiometer Setting	R	y_o	y_m	θ_1
0 Ω	1000 Ω			
2000 Ω	3000 Ω			
4000 Ω	5000 Ω			

Then compare the measured values of θ_1 obtained in Table 7.3 with those in Table 7.7 by completing Table 7.8, where

$$\% \text{ Difference} = \frac{|\theta_1 \text{ (Table 7.3)} - \theta_1 \text{ (Table 7.7)}|}{\theta_1 \text{ (Table 7.3)}} \times 100\% \qquad (7.13)$$

TABLE 7.8

R	θ_1 (Table 7.3)	θ_1 (Table 7.7)	% Difference
1000 Ω			
3000 Ω			
5000 Ω			

Which method provided the higher degree of accuracy for determining θ_1?

Series Sinusoidal Circuits

OBJECT

*To investigate the behavior of series sinusoidal
ac circuits at a fixed frequency.*

EQUIPMENT REQUIRED

Resistors

1 — 1-kΩ

Inductors

1 — 10-mH

Capacitors

1 — 0.01-μF

Instruments

1 — DMM

1 — Oscilloscope

1 — Audio oscillator or function generator

EQUIPMENT ISSUED

TABLE 8.1

Item	Manufacturer and Model No.	Laboratory Serial No.
DMM		
Oscilloscope		
Audio oscillator or function generator		

TABLE 8.2

Resistors	
Nominal Value	Measured Value
18 kΩ	

RÉSUMÉ OF THEORY

Kirchhoff's voltage law is applicable to ac circuits, but it is now stated as follows: *The phasor sum of the voltages around a closed loop is equal to zero.*

For example, in a series circuit, the source voltage is the phasor sum of the component (load) voltages. For a series *R-L-C* circuit,

$$E^2 = V_R^2 + (V_L - V_C)^2$$

so that

$$\boxed{E = \sqrt{V_R^2 + (V_L - V_C)^2}} \tag{8.1}$$

where E is the source voltage, V_R the voltage across the total resistance of the circuit, V_L the voltage across the total inductance, and V_C the voltage across the total capacitance.

Since the resistance and the reactance of a series *R-L-C* circuit are in quadrature,

$$Z^2 = R^2 + X_T^2 \tag{8.2}$$

where X_T (total reactance) $= X_L - X_C$.

In a series *R-L-C* sinusoidal circuit, the voltage across a reactive component may be greater than the input voltage.

The average power (in watts) delivered to a sinusoidal circuit is

$$\boxed{P = I^2 R = VI \cos \theta} \tag{8.3}$$

where the values of the current and the voltage are the effective values.

The power factor F_p is given by

$$\boxed{F_p = \frac{R_T}{Z_T} = \cos\theta} \tag{8.4}$$

where θ is the angle associated with the total impedance Z_T.

For an ideal inductor, the current lags the voltage across it by 90°. For a capacitor, the current leads the voltage across it by 90°. Inductive circuits are therefore called *lagging power-factor circuits,* and capacitive circuits are called *leading power-factor circuits.*

In a purely resistive circuit, the voltage and the current are in phase and the power factor is unity.

PROCEDURE

Part 1   Series *R-L* Circuit

(a) Construct the network of Fig. 8.1(a). Insert the measured values of R and R_l.

FIG. 8.1   (a)

(b) After setting E to $8\,\text{V}\,(p\text{-}p)$, measure the voltage $V_{R(p\text{-}p)}$ with the oscilloscope. Reverse the positions of R and L and measure $V_{L(p\text{-}p)}$ [to ensure that the oscillator (or generator) and the oscilloscope have a common ground].

$V_{R(p\text{-}p)} = $ _____ , $V_{L(p\text{-}p)} = $ _____

(c) Determine $I_{p\text{-}p}$ from $I_{p\text{-}p} = V_{R(p\text{-}p)}/R_{\text{measured}}$.

$I_{p\text{-}p} = $ _____

(d) Determine the input impedance from $Z_T = E_{p\text{-}p}/I_{p\text{-}p}$. Ignore the effects of R_l since $R \gg R_l$ and $X_L \gg R_l$ at $f = 10$ kHz. In other words, consider the coil to be purely ideal at this frequency ($R_l \cong 0\ \Omega$).

$Z_T = $ _____

(e) Calculate the total impedance (magnitude and angle) and current $I_{p\text{-}p}$ from the nameplate value of the inductance and the measured resistor value. Ignore the effect of R_l.

$\mathbf{Z}_T$(theoretical) = _____ , $I_{p\text{-}p}$(calculated) = _____

(f) Compare the results of (d) and (e) and explain the source of any differences.

(g) Show that your measurements from (b) satisfy Kirchhoff's voltage law. That is, show that $E = \sqrt{V_R^2 + V_L^2}$ using peak-to-peak values.

(h) Determine the phase angle between **E** and $\mathbf{V}_R$ using the connections shown in Fig. 8.1 and the dual-trace method introduced in Experiment 7.

$D_1 = $ _____ , $D_2 = $ _____

$\theta_1 = $ _____

(i) Interchange the positions of the resistor and the coil and repeat (h). The phase angle is now between **E** and $\mathbf{V}_L$.

$D_1 = $ _____ , $D_2 = $ _____

$\theta_2 = $ _____

(j) The phase angle between $\mathbf{V}_R$ and $\mathbf{V}_L$ should be 90°. Does $\theta_1 + \theta_2 = 90°$? What is the percent difference $[|(90° - \text{sum})|/90° \times 100\%]$?

% difference = _____

(k) Using $E = 0.707(8/2)$ V $\angle 0° = 2.828$ V $\angle 0°$, $R = 1$ kΩ, and $L = 10$ mH ($R_l = 0$ Ω), *calculate* V_R, V_L, and I at $f = 10$ kHz and draw the phasor diagram. Determine the phase angle θ_1 between E and V_R and compare to the measured value of (h). Determine θ_2 from the phasor diagram and compare to θ_2 of (i).

θ_1(calculated) = _____ , θ_1[from (h)] = _____

θ_2(calculated) = _____ , θ_2[from (i)] = _____

(l) Continue the mathematical analysis to find all the theoretical (second column) values of Table 8.3.

(m) Use the measured values to determine all the quantities of the first column Table 8.3.

TABLE 8.3

Quantity	Measured (or Calculated from Measured Values)	Theoretical (Calculated)
$E_{p\text{-}p}$		
$V_{R(p\text{-}p)}$		
$V_{L(p\text{-}p)}$		
$I_{p\text{-}p}$		
E_{rms}		
$V_{R(rms)}$		
$V_{L(rms)}$		
I_{rms}		
Z_T		
θ_T		
P_T		
F_p		

(n) Compare the measured and theoretical results.

Part 2 Series *R-C* Circuit

(a) Construct the network of Fig. 8.2(a). Insert the measured resistor value.

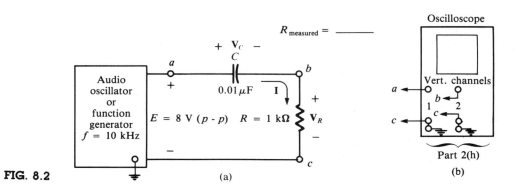

FIG. 8.2

(a)

(b)

(b) After setting E to 8 V $(p\text{-}p)$, measure the voltage $V_{R(p\text{-}p)}$ with the oscilloscope. Reverse the positions of R and C and measure $V_{C(p\text{-}p)}$ [to ensure that the oscillator (or generator) and the oscilloscope have a common ground].

$$V_{R(p\text{-}p)} = \underline{\hspace{2cm}} , \quad V_{C(p\text{-}p)} = \underline{\hspace{2cm}}$$

(c) Determine $I_{p\text{-}p}$ from $I_{p\text{-}p} = V_{R(p\text{-}p)}/R_{\text{measured}}$.

$$I_{p\text{-}p} = \underline{\hspace{2cm}}$$

(d) Determine the input impedance from $Z_T = E_{p\text{-}p}/I_{p\text{-}p}$.

$$Z_T(\text{measured}) = \underline{\hspace{2cm}}$$

(e) Calculate the total impedance (magnitude and angle) and current $I_{p\text{-}p}$ from the nameplate value of the capacitance and the measured resistor value.

$$\mathbf{Z}_T(\text{theoretical}) = \underline{\hspace{2cm}} , \quad I_{p\text{-}p}(\text{calculated}) = \underline{\hspace{2cm}}$$

(f) Compare the results of (d) and (e) and explain the source of any differences.

(g) Show that your measurements from (b) satisfy Kirchhoff's voltage law. That is, show that $E = \sqrt{V_R^2 + V_C^2}$ using peak-to-peak values.

(h) Determine the phase angle between $\mathbf{E}$ and $\mathbf{V}_R$ using the connections shown in Fig. 8.2 and the dual-trace method introduced in Experiment 7.

$D_1 = $ _____ , $D_2 = $ _____

$\theta_1 = $ _____

(i) Interchange the positions of the resistor and capacitor and repeat (h). The phase angle is now between $\mathbf{E}$ and $\mathbf{V}_C$.

$D_1 = $ _____ , $D_2 = $ _____

$\theta_2 = $ _____

(j) The phase angle between $\mathbf{V}_R$ and $\mathbf{V}_C$ should be 90°. Does $\theta_1 + \theta_2 = 90°$? What is the percent difference $[|90° - \text{sum}|/90° \times 100\%]$?

% difference = _____

(k) Using $\mathbf{E} = 0.707(8/2)$ V $\angle 0° = 2.828$ V $\angle 0°$, $R = 1$ kΩ, and $C = 0.01$ μF, *calculate* $\mathbf{V}_R$, $\mathbf{V}_L$, and $\mathbf{I}$ at $f = 10$ kHz and draw the phasor diagram. Determine the phase angle θ_1 between $\mathbf{E}$ and $\mathbf{V}_R$ and compare to the measured value of (h). Determine θ_2 from the phasor diagram and compare to θ_2 of (i).

θ_1 (calculated) = _____ , θ_1 [from (h)] = _____

θ_2 (calculated) = _____ , θ_2 [from (i)] = _____

(l) Continue the mathematical analysis to find all the theoretical (column 2) values of Table 8.4.

(m) Use the measured (column 1) values to determine all the quantities of Table 8.4.

TABLE 8.4

Quantity	Measured (or Calculated from Measured Values)	Theoretical (Calculated)
$E_{p\text{-}p}$		
$V_{R(p\text{-}p)}$		
$V_{C(p\text{-}p)}$		
$I_{p\text{-}p}$		
E_{rms}		
$V_{R(rms)}$		
$V_{C(rms)}$		
I_{rms}		
Z_T		
θ_T		
P_T		
F_p		

(n) Compare the measured and theoretical results.

Part 3   *R-L-C* Network

(a) Construct the network of Fig. 8.3. Insert the measured resistance values. Ignore the effects of R_l in the following analysis.

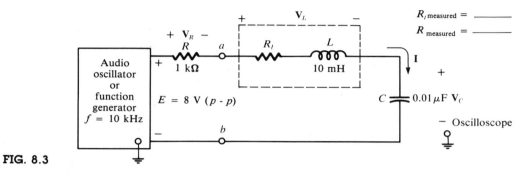

FIG. 8.3

(b) Measure all the component voltages with $E = 8$ V (p-p). To ensure a common ground between the oscilloscope and oscillator for each measurement, make sure the element is placed in the position of the capacitor C. In other words, reverse the order of the elements so that the element across which the voltage is to be measured has the position of C indicated in Fig. 8.3.

$$V_{R(p\text{-}p)} = \underline{\hspace{2cm}} , V_{L(p\text{-}p)} = \underline{\hspace{2cm}} ,$$
$$V_{C(p\text{-}p)} = \underline{\hspace{2cm}}$$

(c) Determine I_{p-p} from $I_{p\text{-}p} = V_{R(p\text{-}p)}/R_{\text{measured}}$.

$$I_{p\text{-}p} = \underline{\hspace{2cm}}$$

(d) Calculate Z_T from $Z_T = E_{p\text{-}p}/I_{p\text{-}p}$.

$$Z_T = \underline{\hspace{2cm}}$$

(e) Using the nameplate values, calculate Z_T and compare to the result of (d).

Z_T(calculated) = _____

(f) Verify Kirchhoff's law by showing that $E = \sqrt{V_R^2 + (V_L - V_C)^2}$ using peak-to-peak values. Draw the phasor diagram using $E = E\angle 0°$ and effective values for the voltages.

(g) Use the voltage divider rule to determine the voltage $V_{ab(p-p)}$.

$V_{ab(p-p)}$(calculated) = _____

(h) Measure the voltage V_{ab} and compare to the result of (g).

$V_{ab(p-p)}$ (measured) = _____

Parallel Sinusoidal Circuits

OBJECT

To investigate the behavior of parallel sinusoidal ac circuits.

EQUIPMENT REQUIRED

Resistors

1 — 1-kΩ
3 — 10-Ω

Inductors

1 — 10-mH

Capacitors

1 — 0.01-μF

Instruments

1 — DMM
1 — Oscilloscope
1 — Audio oscillator or function generator

EQUIPMENT ISSUED

TABLE 9.1

Item	Manufacturer and Model No.	Laboratory Serial No.
DMM		
Oscilloscope		
Audio oscillator or function generator		

TABLE 9.2

Resistors	
Nominal Value	Measured Value
1 kΩ	
10 Ω	
10 Ω	
10 Ω	

RÉSUMÉ OF THEORY

Kirchhoff's current law as applied to ac circuits states that the phasor sum of the currents entering and leaving a node must equal zero. For example, in a parallel circuit, the source current is the phasor sum of the component (load) currents. For a parallel R-L-C circuit, the magnitude of the source current I is given by $I = \sqrt{I_R^2 + I_{X_T}^2}$, where $I_{X_T} = I_{X_L} - I_{X_C}$ (I_{X_L} is the current in the inductive branch, I_{X_C} is the current in the capacitive branch, and I_R is the current in the resistive branch).

In a parallel R-L-C circuit, it is possible for the magnitude of the source current I to be less than a branch current.

Impedances in parallel combine according to the following equation:

$$\frac{1}{Z_T} = \frac{1}{Z_1} + \frac{1}{Z_2} + \frac{1}{Z_3} \qquad (9.1)$$

For two impedances in parallel, it is usually more convenient to use the equation in the form

$$Z_T = \frac{Z_1 Z_2}{Z_1 + Z_2} \qquad (9.2)$$

In parallel R-L-C circuits, it is possible for the total impedance to be larger than a branch impedance.

To obtain the waveform of a current on an oscilloscope, it is necessary to view the voltage across a resistor. Since the voltage and current of a resistor are related by $E = IR$ (a linear relationship), the current and voltage waveforms are always of the same appearance and in phase.

PROCEDURE

Part 1 *R-L Parallel Network*

(a) Construct the network of Fig. 9.1. Insert the measured values of each resistor. The magnitude of R and X_L at the applied frequency permits ignoring (on an approximate basis) the effects of the two sensing resistors, R_{S_1} and R_{S_2}, and the resistance of the inductance R_l when we analyze the system. In other words, assume that you have an ideal parallel *R-L* system.

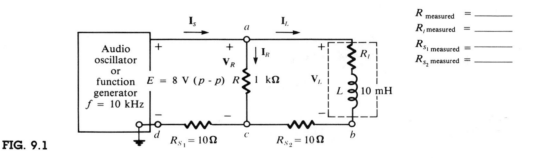

R measured = _____
R_l measured = _____
R_{S_1} measured = _____
R_{S_2} measured = _____

FIG. 9.1

(b) Using the nameplate inductor value (10 mH) and the measured resistance level for R, calculate the various currents of the network ignoring the effect of R_{S_1}, R_{S_2}, and R_l. Note below that the peak-to-peak values are requested, so use peak-to-peak values throughout the calculations.

$$I_{S(p-p)} = \text{_____} , \quad I_{R(p-p)} = \text{_____} ,$$

$$I_{L(p-p)} = \text{_____}$$

Calculate the rms values of the above quantities and insert below.

$$I_{S(rms)} = \underline{\hspace{3cm}}, \quad I_{R(rms)} = \underline{\hspace{3cm}},$$

$$I_{L(rms)} = \underline{\hspace{3cm}}$$

(c) Energize the network of Fig. 9.1 and set the input voltage to 8 V (*p-p*) using the oscilloscope. Use the oscilloscope to measure the voltage $V_{R_{S_1}}$ and insert the value below.

$$V_{R_{S1(p-p)}} = \underline{\hspace{3cm}}$$

(d) To ensure that the oscilloscope and oscillator have a common ground, remove R_{S_1} and connect point *c* to point *d*. Measure the voltages $V_{R_{S_2}}$ and V_R.

$$V_{RS2(p-p)} = \underline{\hspace{3cm}}, \quad V_{R(p-p)} = \underline{\hspace{3cm}}$$

(e) Using the results of (c) and (d), calculate the peak-to-peak values of the various currents of the network using Ohm's law.

$$I_{S(p-p)} = \underline{\hspace{3cm}}, \quad I_{R(p-p)} = \underline{\hspace{3cm}}$$

$$I_{L(p-p)} = \underline{\hspace{3cm}}$$

(f) Calculate the rms values of the measured quantities and insert below. Compare to the results of (b).

$I_{S(rms)} =$ _____ , $I_{R(rms)} =$ _____ ,

$I_{L(rms)} =$ _____

(g) Is the source current I_S larger in magnitude than each branch current? Should it be?

(h) Using the input voltage as a reference ($\mathbf{E} = E \angle 0°$) and the measured rms values of the currents $\mathbf{I}_R$ and $\mathbf{I}_L$, draw a phasor diagram to scale and measure the current $\mathbf{I}_S$. How does it compare to the measured value? From the phasor diagram, determine the angle θ_S between $\mathbf{I}_S$ and $\mathbf{I}_R$ (same as between $\mathbf{E}$ and $\mathbf{I}_S$), the angle θ between $\mathbf{I}_S$ and $\mathbf{I}_L$, and the angle θ_1 between $\mathbf{I}_R$ and $\mathbf{I}_L$.

$I_S(\text{diagram}) =$ _____ , $I_S(\text{measured}) =$ _____

$\theta_S =$ _____ , $\theta =$ _____ , $\theta_1 =$ _____

(i) Connect the oscilloscope as shown in Fig. 9.2. *Do not reintroduce the resistor R_{S_1}.* It will permit a measurement of the phase angle between the two branch currents I_R and I_L. Record the value of θ_1. Points a, b, and c appear on Fig. 9.1. Use the dual-trace method to determine the phase angle.

Oscilloscope

a — 1   2 — b

c — 1   2 — c

FIG. 9.2   Vertical channels

θ_1(measured) = _____ , θ_1[from (h)] = _____

(j) Reintroduce the resistor R_{S_1} and connect the oscilloscope as shown in Fig. 9.3. The phase angle obtained will be between the input voltage $\mathbf{E}$ and the current $\mathbf{I}_S$, or between $\mathbf{I}_R$ and $\mathbf{I}_S$, since $\mathbf{E}$ and $\mathbf{I}_R$ are in phase.

Oscilloscope

a — 1   2 — c

d — 1   2 — d

FIG. 9.3   Vertical channels

θ_S(measured) = _____ , θ_S[from (h)] = _____

(k) Using the fact that $\theta_S + \theta + \theta_1 = 180°$ for the right triangle of (h), find the measured value of θ using the measured values of (i) and (j). Compare to the result of (h).

$$\theta \text{(measured)} = \text{_____} , \theta \text{[from (h)]} = \text{_____}$$

Part 2   *R-C* Parallel Network

(a) Construct the network of Fig. 9.4. Insert the measured values of the resistors. As in Part 1, ignore the magnitude of R_{S_1} and R_{S_2} as compared to the other impedances of the network when making your calculations.

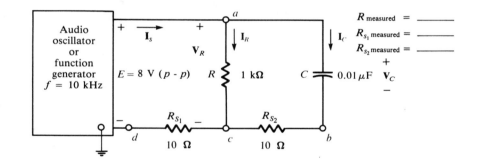

FIG. 9.4

(b) Using the nameplate capacitor value (0.01 μF) and the measured resistance level for R, calculate the currents indicated below.

$I_{S(p\text{-}p)} =$ ——————— , $I_{R(p\text{-}p)} =$ ——————— ,

$I_{C(p\text{-}p)} =$ ———————

Calculate the rms values of the above quantities and insert below.

$I_{S(\text{rms})} =$ ——————— , $I_{R(\text{rms})} =$ ——————— ,

$I_{C(\text{rms})} =$ ———————

(c) Energize the network of Fig. 9.4 and set the input voltage to 8 V ($p\text{-}p$) using the oscilloscope. Use the oscilloscope to measure the voltage $V_{R_{S_1}}$ and insert the value below.

$V_{R_{S1(p\text{-}p)}} =$ ———————

Calculate the rms value of $V_{R_{S_1}}$.

$V_{R_{S1(\text{rms})}} =$ ———————

(d) To ensure that the oscilloscope and oscillator have a common ground, remove R_{S_1} and connect point c to point d. Measure the voltages $V_{R_{S_2}}$ and V_R.

$$V_{R_{S2(p\text{-}p)}} = \underline{\hspace{3cm}} , V_{R(p\text{-}p)} = \underline{\hspace{3cm}}$$

Calculate the rms values of $V_{R_{S_2}}$ and V_R.

$$V_{R_{S2(rms)}} = \underline{\hspace{3cm}} , V_{R(rms)} = \underline{\hspace{3cm}}$$

(e) Using the results of (c) and (d), calculate the peak-to-peak values of the various currents of the network using Ohm's law.

$$I_{S(p\text{-}p)} = \underline{\hspace{3cm}} , I_{R(p\text{-}p)} = \underline{\hspace{3cm}} ,$$

$$I_{C(p\text{-}p)} = \underline{\hspace{3cm}}$$

(f) Calculate the rms values of the measured quantities and insert below. Compare to the results of (b).

$$I_{S(rms)} = \underline{\hspace{3cm}} , I_{R(rms)} = \underline{\hspace{3cm}} ,$$

$$I_{C(rms)} = \underline{\hspace{3cm}}$$

(g) Is the source current I_S larger in magnitude than each branch current? Should it be?

(h) Using the input voltage as a reference ($\mathbf{E} = E \angle 0°$) and the measured rms values of the currents $\mathbf{I}_R$ and $\mathbf{I}_C$, draw a phasor diagram to scale and measure the current $\mathbf{I}_S$. How does it compare to the measured value? From the phasor diagram, determine the angle θ_S between $\mathbf{I}_S$ and $\mathbf{I}_R$ (same as between $\mathbf{E}$ and $\mathbf{I}_S$), the angle θ between $\mathbf{I}_S$ and $\mathbf{I}_C$, and the angle θ_1 between $\mathbf{I}_R$ and $\mathbf{I}_C$.

I_S(diagram) = _____ , I_S(measured) = _____

θ_S = _____ , θ = _____ , θ_1 = _____

(i) Connect the oscilloscope as shown in Fig. 9.2. *Do not reintroduce the resistor R_{S_1}.* It will permit a measurement of the phase angle between the two branch currents I_R and I_C. Record the value of θ_1. Points *a*, *b*, and *c* appear in Fig. 9.4. Use the dual-trace method to determine the phase angle.

θ_1(measured) = _____ , θ_1[from (h)] = _____

(j) Reintroduce the resistor R_{S_1} and connect the oscilloscope as shown in Fig. 9.3. The phase angle obtained will be between the input voltage E and current I_S, or between I_R and I_S, since E and I_R are in phase.

θ_S(measured) = _____ , θ_S[from (h)] = _____

(k) Using the fact that $\theta_S + \theta + \theta_1 = 180°$, find the measured value of θ using the measured values of (i) and (j). Compare to the result of (h).

$\theta(\text{measured}) = \underline{\hspace{3cm}}$, $\theta[\text{from (h)}] = \underline{\hspace{3cm}}$

Part 3   *R-L-C Parallel Network*

(a) Construct the network of Fig. 9.5. Insert the measured resistor values. As in the earlier parts, ignore the sensing resistors and R_l compared to R, X_L, and X_C in your calculations.

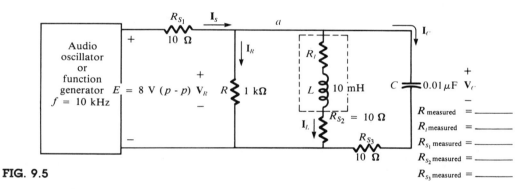

FIG. 9.5

R measured $= \underline{\hspace{1.5cm}}$
R_l measured $= \underline{\hspace{1.5cm}}$
R_{S_1} measured $= \underline{\hspace{1.5cm}}$
R_{S_2} measured $= \underline{\hspace{1.5cm}}$
R_{S_3} measured $= \underline{\hspace{1.5cm}}$

(b) Measure the peak-to-peak values of V_R, V_{RS2}, and V_{RS3} and determine their rms values.

$V_{R(p\text{-}p)} = \underline{\hspace{3cm}}$, $V_{RS2(p\text{-}p)} = \underline{\hspace{3cm}}$,

$V_{RS3(p\text{-}p)} = \underline{\hspace{3cm}}$

$V_{R(rms)} = \underline{\hspace{3cm}}$, $V_{RS2(rms)} = \underline{\hspace{3cm}}$,

$V_{RS3(rms)} = \underline{\hspace{3cm}}$

(c) Reverse the input leads to the oscillator and measure the voltage $V_{R_{S_1}}$.

$V_{R_{S_1(p\text{-}p)}} = \underline{\hspace{3cm}}$, $V_{R_{S_1(rms)}} = \underline{\hspace{3cm}}$

(d) Verify Kirchhoff's current law at node a. That is, show with a phasor diagram (with $\mathbf{E}$ as the reference: $\mathbf{E} = E \angle 0°$) that the vector sum of $\mathbf{I}_R$, $\mathbf{I}_L$, and $\mathbf{I}_C$ is equal to $\mathbf{I}_S$. The currents can be determined from the measurements of (b) and (c).

(e) Using the nameplate values, calculate the source current $\mathbf{I}_S$ and compare to the value obtained in (d).

$\mathbf{I}_{S(rms)}$(measured) = $\underline{\hspace{3cm}}$, $\mathbf{I}_{T(rms)}$(calculated) = $\underline{\hspace{3cm}}$

Series-Parallel Sinusoidal Circuits

OBJECT

To investigate the behavior of series-parallel sinusoidal networks.

EQUIPMENT REQUIRED

Resistors

1 — 470-Ω, 1-kΩ

Inductors

1 — 10-mH

Capacitors

1 — 0.02-μF

Instruments

1 — DMM
1 — Oscilloscope
1 — Audio oscillator or function generator

EQUIPMENT ISSUED

TABLE 10.1

Item	Manufacturer and Model No.	Laboratory Serial No.
DMM		
Oscilloscope		
Audio oscillator or function generator		

TABLE 10.2

Resistors	
Nominal Value	Measured Value
470 Ω	
1 kΩ	

RÉSUMÉ OF THEORY

In the previous experiments, we showed that Kirchhoff's voltage and current laws hold for ac series and parallel circuits. In fact, all of the previously used rules, laws, and methods of analysis apply equally well for both dc and ac networks. The major difference for ac circuits is that we now use reactance, resistance, and impedance instead of solely resistance.

Consider the common series-parallel circuit of Fig. 10.1.

FIG. 10.1

The various impedances shown could be made up of many elements in a variety of configurations. No matter how varied or numerous the elements might be, Z_1, Z_2, and Z_3 represent the total impedance for that branch. For example, Z_1 might be as shown in Fig. 10.2.

FIG. 10.2

The currents and voltages in Fig. 10.1 can be found by applying any of the methods outlined for dc networks; in particular,

$$\mathbf{Z}_T = \mathbf{Z}_1 + \mathbf{Z}_2 \| \mathbf{Z}_3 \qquad \mathbf{I}_S = \mathbf{I}_1 = \frac{\mathbf{E}}{\mathbf{Z}_T}$$

and

$$\mathbf{I}_2 = \frac{\mathbf{Z}_3 \mathbf{I}_S}{\mathbf{Z}_3 + \mathbf{Z}_2} \qquad \text{(current divider rule)}$$

$\mathbf{I}_3$ can be determined using Kirchhoff's current law:

$$\mathbf{I}_3 = \mathbf{I}_S - \mathbf{I}_2$$

Ohm's Law provides the voltages:

$$\mathbf{V}_1 = \mathbf{I}_1 \mathbf{Z}_1 \qquad \mathbf{V}_2 = \mathbf{I}_2 \mathbf{Z}_2 \qquad \mathbf{V}_3 = \mathbf{I}_3 \mathbf{Z}_3$$

Keep in mind that the voltages and currents are phasor quantities that have both magnitude and an associated angle.

The power factor $\cos \theta = R_T/Z_T$, where R_T is the equivalent total resistance of the circuit at the input terminals.

PROCEDURE

Part 1 *R-L Series-Parallel Network*

(a) Construct the network of Fig. 10.3. Insert the measured value for each resistor. At the applied frequency, the resistor R_l can be ignored in comparison with the other elements of the network. Set the source to 8 V (*p-p*) with the oscilloscope.

R_1 measured = 471 Ω
R_2 measured = 1010 Ω
R_l measured = 13 Ω

FIG. 10.3

(b) Calculate the total impedance of the network at $f = 10$ kHz. Use measured resistor values.

$$Z_T = \underline{877.5 \angle 30.89°}$$

(c) Using the results of part 1(b), calculate the peak-to-peak value of the source current $I_S = I_1$.

$$I_{S(p-p)} = \underline{3.22mA \angle -30.89°}$$

(d) Measure the voltage V_1 with the DMM. Convert the reading to a peak-to-peak value and calculate the peak-to-peak value of $\mathbf{I_1} = \mathbf{I_S}$.

$$V_{1(rms)} \text{ (measured)} = \underline{1.59v}$$
$$V_{1(p-p)} = \underline{4.497}$$
$$I_{1(p-p)} = I_{S(p-p)} = \underline{9.569mA}$$
$$3.383 mA(rms)$$

How does the magnitude of $I_{S(p-p)}$ compare with the results of part 1(c)?

(e) Measure the peak-to-peak values of $\mathbf{V}_2$ and $\mathbf{V}_L$ with the oscilloscope by connecting the ground terminal of the scope channel to point b and the high end of the channel to point a.

$V_{2(p-p)} = $ _4.865V_

1.72 RMS $V_{L(p-p)} = $ _4.865V_

(f) Using the results of part 1(e), calculate the peak-to-peak values of $\mathbf{I}_2$ and $\mathbf{I}_L$.

$I_{2(p-p)} = $ _4.865_

$I_{L(p-p)} = $ _7.743_

(g) Using the results of parts 1(d) and 1(f), determine whether the following relationship is satisfied using peak-to-peak values:

$$I_1 = \sqrt{I_2^2 + I_L^2}$$

$$\sqrt{4.865^2 + 7.743^2} = 9.145$$

(h) Devise a method to measure the phase angle between $\mathbf{E}$ and $\mathbf{I}_S = \mathbf{I}_1$. Check your method with the instructor and then proceed to determine the angle.

$\theta_T = $ _-28.8_

(i) Devise a method to measure the phase angle between $\mathbf{E}$ and $\mathbf{V}_2 (= \mathbf{V}_L)$. Check your method with the instructor and then proceed to determine the angle.

$\theta = \underline{25.2}$

(j) Using the results of part 1(h), calculate the phase angle between $\mathbf{E}$ and $\mathbf{V}_1$.

$\theta = \underline{\hspace{3cm}}$

(k) Draw the phasor diagram of $\mathbf{E}$, $\mathbf{V}_1$, $\mathbf{V}_2$, and $\mathbf{I}_S$ using the above results and peak-to-peak values.

Part 2   R-C Series-Parallel Network

(a) Construct the network of Fig. 10.4. Insert the measured resistor values. Set the source to 8 V (*p-p*) with the oscilloscope.

FIG. 10.4

(b) Calculate the magnitude of the total impedance of the network at $f = 10$ kHz.

$Z_T = $ <u>$988.34 \angle -29.54$</u>

(c) Measure the rms value of V_1 with the DMM.

$V_{1(rms)} = $ <u>1.29 rms</u>

Calculate the peak-to-peak value of V_1.

$V_{1(p-p)} = $ <u>3.649 p-p</u>

Calculate the magnitude of Z_T from

$$Z_T = \frac{E_{p-p}}{I_{S(p-p)}}$$

using the fact that $I_{S(p-p)} = I_{1(p-p)} = [V_{1(p-p)}/R_1]$

$I_{S(p-p)} = $ <u>7.764 mA</u>
$Z_T = $ <u>1030.4</u>

How do the results for Z_T from parts 2(b) and 2(c) compare?

(d) Measure the peak-to-peak values of $\mathbf{V}_2$ and $\mathbf{V}_C$ with the oscilloscope.

$V_{2(p-p)} = $ <u>$(1.83$ rms$)$</u> 5.176
$V_{C(p-p)} = $ _____

Calculate the peak-to-peak values of I_2 and I_C.

$$I_{2(p-p)} = \underline{5.18 mA}$$
$$I_{C(p-p)} = \underline{6.50 mA}$$

(e) Using the results of parts 2(c) and 2(d), is the following relationship satisfied? Use peak-to-peak values.

$$I_1 = \sqrt{I_2^2 + I_C^2}$$

(f) Devise a method to measure the phase angle between **E** and **I**$_S$. Check the method with your instructor and then proceed to determine the angle.

$$\theta_T = \underline{28.8}$$

(g) Devise a method to measure the phase angle between **E** and **I**$_2$. Check the method with your instructor and then proceed to determine the angle.

$$\theta = \underline{-19.8}$$

(h) Devise a method to determine the phase angle between **I**$_2$ and **I**$_C$. Check the method with your instructor and then proceed to determine the angle.

$\theta =$ _____

(i) If $\mathbf{E} = E \angle 0°$, write $\mathbf{V}_1$ and $\mathbf{V}_2$ in phasor form using the results obtained above. Use peak-to-peak values for the magnitudes.

$\mathbf{V}_1 =$ _____

$\mathbf{V}_2 =$ _____

(j) Is the following Kirchhoff's voltage law relationship satisfied? Use effective values.

$$E = \sqrt{V_1^2 + V_2^2}$$

Step 1: Remove R. See Fig. 11.2(a).
Step 2: Replace **E** by a short-circuit equivalent. See Fig. 11.2(a).
Step 3: Solve for $\mathbf{Z}_{Th}$. In this case, $\mathbf{Z}_{Th} = \mathbf{Z}_1 || \mathbf{Z}_2$. See Fig. 11.2(a).
Step 4: Replace **E**. See Fig. 11.2(b).
Step 5: Calculate $\mathbf{E}_{Th}$ from

$$\mathbf{E}_{Th} = \frac{\mathbf{Z}_2}{\mathbf{Z}_1 + \mathbf{Z}_2}\,\mathbf{E}$$

See Fig. 11.2(b).
Step 6: Draw the equivalent circuit and replace R. See Fig. 11.2(c).
Step 7: Calculate **I** from

$$\mathbf{I} = \frac{\mathbf{E}_{Th}}{\mathbf{Z}_{Th} + \mathbf{R}}$$

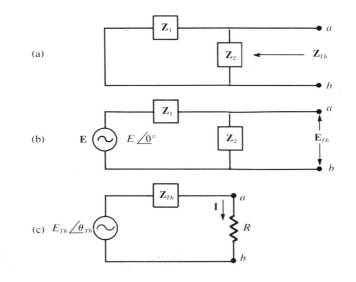

FIG. 11.2

MAXIMUM POWER TRANSFER THEOREM

The maximum power transfer theorem states that for circuits with ac sources, maximum power will be transferred to a load when the load impedance is the conjugate of the Thevenin impedance across its terminals. See Fig. 11.3.

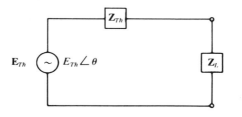

FIG. 11.3

For maximum power transfer, if

$$\mathbf{Z}_{Th} = |Z_{Th}| \angle\theta$$

then

$$\mathbf{Z}_L = |Z_{Th}| \angle-\theta$$

PROCEDURE

Part 1

(a) Measure the dc resistance of the coil and include it in Fig. 11.4. Indicate the measured values of R, R_1, and R_2 from Table 11.2.

FIG. 11.4

R_1 measured = 465 Ω
R_2 measured = 1000 Ω
R measured = 2.21 k
R_l measured = 12 Ω

(b) With E set at 8 V (p-p), measure $V_{R(p\text{-}p)}$ and calculate $V_{R(\text{rms})}$.

$$V_{R(p\text{-}p)} = \underline{4.7\,Vp\text{-}p} \quad, \quad V_{R(\text{rms})} = \underline{1.66\,Vrms}$$

(c) Remove the resistor R and measure the open-circuit voltage across the resulting terminals. This is the magnitude of $\mathbf{E}_{Th}$.

$$|\mathbf{E}_{Th}|_{p\text{-}p} = \underline{5.5\,Vp\text{-}p} \quad, \quad |\mathbf{E}_{Th}|_{\text{rms}} = \underline{1.94\,Vrms}$$

(d) Using the measured resistance levels, calculate the magnitude of $\mathbf{E}_{Th}$ and its associated angle. Compare with the magnitude of part 1(c). Assume that $\mathbf{E} = E \angle 0° = 0.707(8/2) \angle 0° = 2.828$ V $\angle 0°$.

$$\mathbf{E}_{Th} = \underline{1.92}$$

(e) Calculate $\mathbf{Z}_{Th}$ (magnitude and angle) using the measured resistor values and the nameplate inductor level of 10 mH. The applied frequency is 10 kHz.

$$\mathbf{Z}_{Th} = \underline{704.7 \angle 63}$$

Determine the resistive and reactive components of $\mathbf{Z}_{Th}$.

$$R = \underline{319.7} \quad, \quad X_L = \underline{628 \angle 90}$$

(f) Insert the calculated values obtained for $\mathbf{E}_{Th}$ and $\mathbf{Z}_{Th}$ in parts 1(c) and 1(e) in Fig. 11.5. Using the 1-kΩ potentiometer and the amplitude control on the audio oscillator or function generator, set the values of the Thevenin circuit. Calculate the inductance level from the results of part 1(e) and the fact that $L = X_L/\omega$. Then measure the voltage $V_{R(p-p)}$ and compare with the measurement of part 1(b).

FIG. 11.5

$$V_{R(p\text{-}p)} = \underline{4.6} \quad , \quad V_{R(\text{rms})} = \underline{1.63v}$$

(g) Is the Thevenin theorem verified?

Yes

Part 2   Maximum Power Transfer

(a) Construct the circuit of Fig. 11.6. Include the measured resistor values.

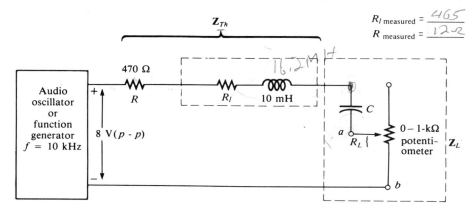

R_l measured = 465
R measured = 12Ω

FIG. 11.6

(b) Set the 0–1-kΩ potentiometer to $R_T = R_{\text{measured}} + R_l$ as required for maximum power transfer. Indicate this value in each row of Table 11.3.

$$R_T = R\,(\text{measured}) + R_l = \underline{465} + \underline{12\Omega} = \underline{477}$$

TABLE 11.3

R_T	X_L	$|Z_{Th}|$	$C\ (\mu F)$	X_C	$|Z_L|$	$V_{ab(\text{rms})}$	$P_L = V_{ab}^2/R_T$
477	628∠90	789∠53	0.0047	3183∠-90	3219∠-81	3.34	23.39
477	628∠90	789∠53	0.01	1592∠-90	1662∠-73	3.46	25.1
477	628∠90	789∠53	0.02	796∠-90	928∠-59	2.71	15.4
477	628∠90	789∠53	0.0247	637∠-90	796∠-53	2.36	
477	628∠90	789∠53	0.05	318∠-90	573∠-34	1.62	4.6 m w
477	628∠90	789∠53	0.1	159∠-90	502∠-18	1.34	4 m w

9.4 p-p
9.6 p-p
8 v p-p
7.2 vp-p
4.2 v p-p
3.9 v p-p

(i) Is maximum power transferred when $R_L = R_T$ and $X_C = X_L$ as required by the theorem?

(j) Assuming that L is exactly 10 mH, what value of capacitance will ensure that $X_L = X_C$ at $f = 10$ kHz? How close do we come to this value in Table 11.3?

Series Resonant Circuits

OBJECT

To investigate the characteristics of a series resonant circuit.

EQUIPMENT REQUIRED

Resistors

1 — 47-Ω, 220-Ω

Inductors

1 — 1-mH, 10-mH

Capacitors

1 — 0.1-μF, 1-μF

Instruments

1 — DMM
1 — Audio oscillator or function generator

EQUIPMENT ISSUED

TABLE 12.1

Item	Manufacturer and Model No.	Laboratory Serial No.
DMM		
Audio oscillator or function generator		

TABLE 12.2

Resistors	
Nominal Value	**Measured Value**
47 Ω	
220 Ω	

RÉSUMÉ OF THEORY

In a series R-L-C circuit, there exists one frequency at which $X_L = X_C$ or $\omega L = 1/\omega C$. At this frequency the circuit is in resonance, and the input voltage and current are in phase. At resonance, the circuit is resistive in nature and has a minimum value of impedance or a maximum value of current.

The resonant radian frequency $\omega_s = 1/\sqrt{LC}$ and frequency $f_s = 1/2\pi\sqrt{LC}$. The Q of the circuit is defined as $\omega_s L/R$ and affects the selectivity of the circuit. High-Q circuits are very selective. At resonance, $V_C = QE_{input}$. The half-power frequencies f_1 and f_2 are defined as those frequencies at which the power dissipated is one-half the power dissipated at resonance. In addition, the current is 0.707 (or $1/\sqrt{2}$) times the current at resonance.

The bandwidth $BW = f_2 - f_1$. The smaller the bandwidth, the more selective the circuit. In Fig. 12.1, note that increasing R results in a less selective circuit. Figure 12.2

FIG. 12.1

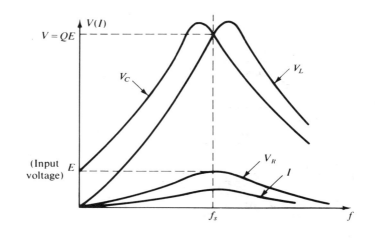

FIG. 12.2

shows the voltages across the three elements versus the frequency. The voltage across the resistor, V_R, has exactly the same shape as the current since it differs by the constant R. V_R is a maximum at resonance. V_C and V_L are equal at resonance (f_s) since $X_L = X_C$, but note that they are not maximum at the resonant frequency. At frequencies below f_s, $V_C > V_L$; at frequencies above f_s, $V_L > V_C$, as indicated in Fig. 12.2.

PROCEDURE

Part 1   Low-Q Circuit

(a) Construct the circuit of Fig. 12.3. Insert the measured resistance values.

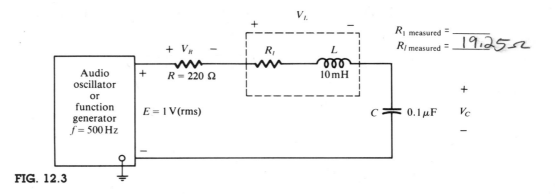

$R_{l \text{ measured}} = $ _____
$R_{l \text{ measured}} = \underline{19.25\,\Omega}$

FIG. 12.3

(b) Using the nameplate values and measured resistance values, compute the radian frequency ω_s, the frequency f_s, and the Q_s of the circuit at resonance. Include the effects of R_l in your calculations.

Res $f_s = 5263.2$
$U_s = 1.994$
$V_R = .843$
$V_L = 1.259$
$V_C = 1.237$

$$\omega_s = \underline{31622.78}, \quad f_s = \underline{5032.92}, \quad Q_s = \underline{1.32}$$

(c) Energize the circuit and set the oscillator to the frequencies indicated in Table 12.3. At each frequency, reset the input to 1 V (rms) with the DMM and measure the rms values of the voltages V_C, V_R, and V_L with the DMM. Take a few extra readings near the resonant frequency to improve your data.

TABLE 12.3

Frequency	$V_{R(rms)}$	$V_{L(rms)}$	$V_{C(rms)}$	$I_{rms} = \dfrac{V_{R(rms)}}{R}$
10 X .2ms 500 Hz	.067 VRms	.011	1.005	
5 X .2ms 1,000 Hz	.134 Vems	.040	1.022	
10x 50u 2,000 Hz	.288 vems	.166	1.100	
6.5X 50u 3,000 Hz	.490	.433	1.227	
5 X 50u 4,000 Hz	.706	.818	1.339	
10 X 20 5,000 Hz	.839	1.216	1.289	
8.3 X 20 6,000 Hz	.802	1.373	1.029	
7.2 X 20u 7,000 Hz	.688	1.353	.780	
6.2 X 20u 8,000 Hz	.599	1.300	.613	
5.5X 20u 9,000 Hz	.502	1.233	.459	
5 X 20u 10,000 Hz	.450	1.193	.374	
f_s (calculated), Hz				
Near $\{$ f_s $\{$				

1.271

Part 2 Higher-Q Circuit

(a) Repeat parts 1(a) through 1(c), replacing the 220-Ω resistor with a resistor of 47 Ω. Record the results in Table 12.4.

$\omega_s =$ _____ , $f_s =$ _____ , $Q_s =$ _____

TABLE 12.4

Frequency	$V_{R(rms)}$	$V_{L(rms)}$	$V_{C(rms)}$	$I_{rms} = \dfrac{V_{R(rms)}}{R}$
500 Hz	.014	.011	1.008	
1,000 Hz	.028	.041	1.034	
2,000 Hz	.065	.175	1.164	
3,000 Hz	.128	.530	1.485	
4,000 Hz	.256	1.361	2.24	
5,000 Hz	.476	3.25	3.21	
6,000 Hz	.344	2.85	2.08	
7,000 Hz	.238	1.988	1.226	
8,000 Hz	.165	1.620	.763	
9,000 Hz	.130	1.435	.550	
10,000 Hz	.109	1.326	.424	
f_s(calculated), Hz				
Near f_s {				
{				

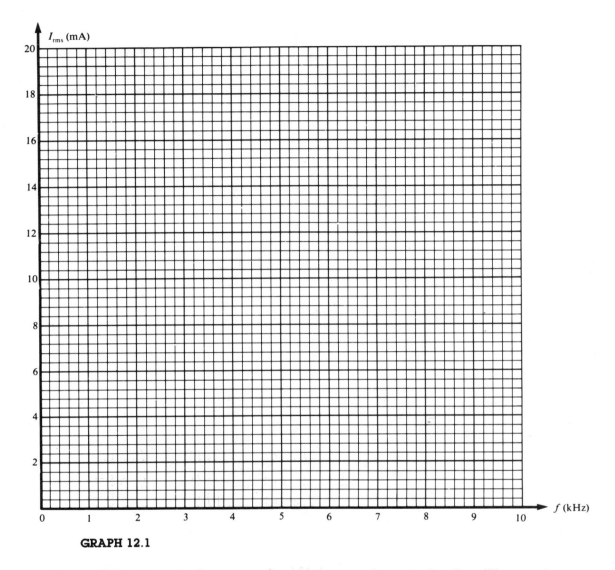

GRAPH 12.1

(b) On Graph 12.1, draw the curves of current I_{rms} (in milliamperes) versus frequency for parts 1 and 2. Label each curve.

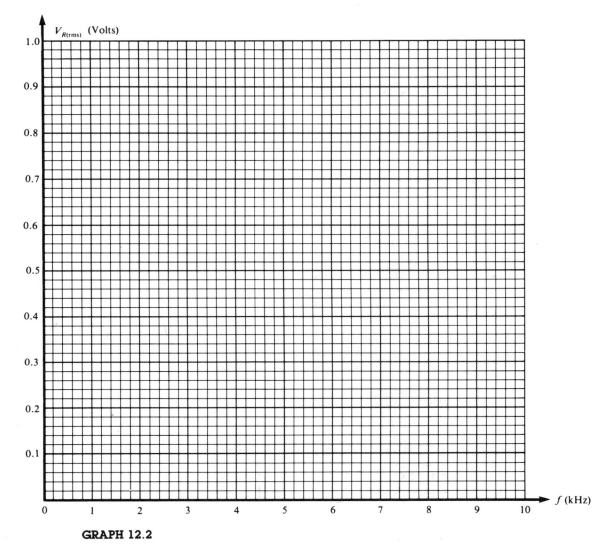

GRAPH 12.2

(c) On Graph 12.2, plot $V_{R(rms)}$ versus frequency for parts 1 and 2. Label each curve.

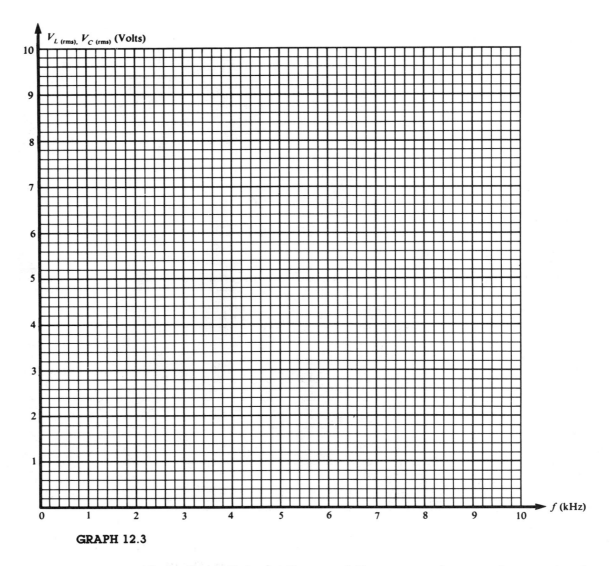

GRAPH 12.3

      (d) On Graph 12.3, plot $V_{L(\text{rms})}$ and $V_{C(\text{rms})}$ versus frequency for parts 1 and 2. Label each curve.

      (e) How do the curves of parts 2(b) through 2(d) compare with those in the Résumé of Theory?

(f) Indicate the resonant and half-power frequencies and the bandwidth on each curve. What is the bandwidth in hertz for each curve using the results of part 2(b)?

BW $(R = 220 \ \Omega) =$ _____ , BW $(R = 47 \ \Omega) =$ _____

Which curve is more selective? Why?

(g) Calculate the Q_s for each part from the data *at the resonant frequency*. That is, use the rms values of V_C and E and the following equation:

$$Q_s = \frac{V_C}{E}$$

$Q_s(R = 220 \ \Omega) =$ _____ , $Q_s(R = 47 \ \Omega) =$ _____

(h) Calculate the Q_s of each circuit from the curves using $Q_s = f_s/\text{BW}$. How do the Q_s's obtained in parts 2(g) and 2(h) compare?

$Q_s(R = 220 \ \Omega) =$ _____ , $Q_s(R = 47 \ \Omega) =$ _____

Part 3

(a) If the inductor of Fig. 12.3 is changed to 1 mH and the capacitance to 1 μF, how will the response curve change?

(b) Make the circuit changes suggested in 3(a) and plot the curve of I_{rms} versus frequency on Graph 12.1. Create your own table below for the data obtained. Does the resulting plot verify the conclusions of 3(a)?

Parallel Resonant Circuits

OBJECT

To investigate the characteristics of a parallel resonant circuit.

EQUIPMENT REQUIRED

Resistors

1 — 33-Ω, 10-kΩ

Inductors

1 — 10-mH

Capacitors

1 — 0.1-μF

Instruments

1 — DMM
1 — Oscilloscope
1 — Audio oscillator or function generator

EQUIPMENT ISSUED

TABLE 13.1

Item	Manufacturer and Model No.	Laboratory Serial No.
DMM		
Oscilloscope		
Audio oscillator or function generator		

TABLE 13.2

Resistors	
Nominal Value	Measured Value
33 Ω	34.4
10 kΩ	9.92

RÉSUMÉ OF THEORY

The basic components of a parallel resonant network appear in Fig. 13.1.

FIG. 13.1

For parallel resonance, the following condition must be satisfied:

$$\frac{1}{X_C} = \frac{X_L}{X_L^2 + R_l^2}$$

(13.1)

At resonance, the impedance of the network is resistive and is determined by

$$Z_{T_p} = \frac{L}{R_l C}$$

(13.2)

The resonant frequency is determined by

$$f_p = \frac{1}{2\pi\sqrt{LC}}\sqrt{1 - \frac{R_l^2 C}{L}}$$

(13.3)

which reduces to

$$f_p = \frac{1}{2\pi\sqrt{LC}}$$

(13.4)

when

$$Q_{coil} = \frac{X_L}{R_l} \geq 10$$

For parallel resonance, the curve of interest is that of V_C versus frequency due to electronic considerations that often place this voltage at the input to the following stage. It has the same shape as the resonance curve for the series configuration with the bandwidth determined by

$$\text{BW} = f_2 - f_1$$

(13.5)

where f_2 and f_1 are the cutoff or band frequencies.

PROCEDURE

Part 1   High-Q Parallel Resonant Circuit

(a) Construct the network of Fig. 13.2. Insert the measured resistance values.

$R_{measured} =$ _____
$R_l\text{ measured} = 11.32$

FIG. 13.2

(b) The input voltage E must be maintained at 10 V (p-p). Make sure to reset E after each new frequency is set on the oscillator.

Since the input to a parallel resonant circuit is usually a constant-current device such as a transistor, the input current I_s has been designed to be fairly constant in magnitude for the frequency range of interest. The input impedance (Z_p) of the parallel resonant circuit will always be sufficiently less than the 10-kΩ resistor, to permit the following approximation:

$$I_s = \frac{E}{10\ k\Omega + Z_p} \cong \frac{E}{10\ k\Omega} = \frac{10\ V\ (p\text{-}p)}{10\ k\Omega} = 1\ mA\ (p\text{-}p)$$

(c) Compute the resonant frequency of the network of Fig. 13.2 using the nameplate data and the measured resistor values.

$f_p =$ _50 29.71_

(d) Calculate the input impedance at resonance using the following equation:

$$Z_p = R_p = \frac{L}{R_l C}$$

$Z_p =$ _8833.92_

(e) Compare the 10-kΩ resistor with the result of part 1(c). Is it reasonable to assume that the input current is fairly constant for the frequency range of interest?

(f) Is this a high- or low-Q circuit? Find Q_p using

$$Q_p = \frac{X_L}{R_l}$$

$Q_p =$ _____

(g) Vary the frequency from 500 Hz to 10 kHz and use the oscilloscope to measure $V_{C(p-p)}$ at each frequency. Make sure to maintain E at 10 V (p-p) for each frequency. Insert in Table 13.3. At each frequency, measure $V_{s(rms)}$ with the DMM and calculate $V_{s(p-p)}$ from $V_{s(p-p)} = 2(1.414 V_{s(rms)}) = 2.828 V_{s(rms)}$. Insert $V_{s(p-p)}$ in Table 13.3. Calculate $I_{s(p-p)}$ from $I_{s(p-p)} = V_{s(p-p)}/R_s$ at each frequency and complete Table 13.3. Take a few extra readings around f_p.

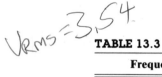

Vrms = 3.54

TABLE 13.3

Frequency	rms $V_{C(p-p)}$	rms $V_{s(p-p)}$	$I_{s(p-p)}$
10x.2ms 500 Hz	.012	3.53	
5x.2ms 1,000 Hz	.022	3.53	
10x 50u 2,000 Hz	.051	3.53	
6.5x50u 3,000 Hz	.098	3.53	
5x 50u 4,000 Hz	.214	3.5	
10x 20u 5,000 Hz	.796	2.90	
8.3x20u 6,000 Hz	.414	3.38	
7.2x20u 7,000 Hz	.201	3.49	
6.2x20u 8,000 Hz	.123	3.52	
5.5x20u 9,000 Hz	.093	3.52	
5x20u 10,000 Hz	.074	3.53	
f_p(calculated), Hz			
Near f_p 5029	.88	2.82	
5263	1.03	2.49	

(h) Plot $V_{C(p-p)}$ versus frequency on Graph 13.1. The choice of vertical scaling is part of the exercise.

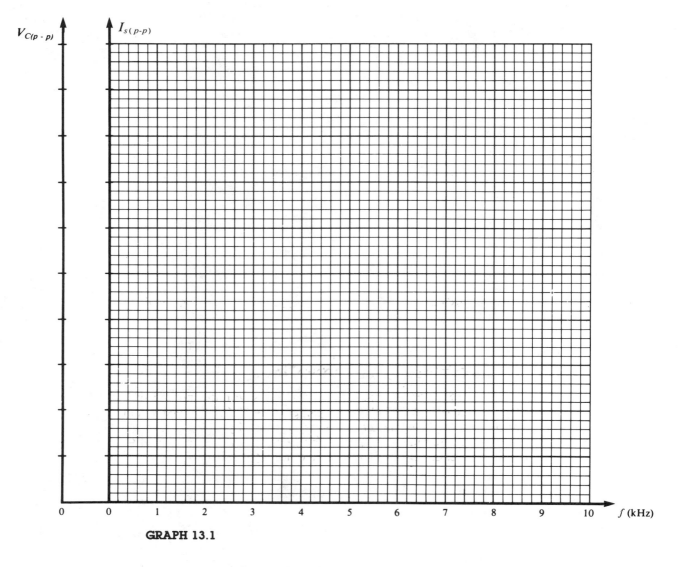

GRAPH 13.1

(i) Plot the curve of $I_{s(p-p)}$ versus f on Graph 13.1. The choice of vertical scaling is left as an exercise.

(j) Is $I_{s(p-p)}$ fairly constant for the frequency range of interest?

(k) Using the results of part 1(h), determine the bandwidth and cutoff frequencies.

BW = _____ , f_1 = _____ , f_2 = _____

(l) Compute the bandwidth from BW = f_p/Q_p and compare with the result of part 1(k).

$BW_{computed}$ = _____ , $BW_{measured}$ [part 1(k)] = _____

Part 2 Lower-Q Parallel Resonant Circuit

(a) Insert the 33-Ω resistor in series with the inductor of Fig. 13.2 so that the total resistance of the inductor branch is $R_l + 33$ Ω (measured value). The effect will be to reduce the Q_p of the network as determined by $Q_p = X_L/R$.

Determine the resonant frequency of the network using Eq. (13.3). The value of R in the equation must now include the 33-Ω resistor.

f_p = _4857._

(b) Calculate the input impedance at resonance.

$$Z_{T_p} = 2083$$

(c) Calculate the Q_p of the network at the resonant frequency. Compare with the result of part 1(f).

$Q_p = $ _____ , Q_p [from part 1(f)] = _____

(d) Repeat part 1(g) and insert the data in Table 13.4. Take a few extra readings around the resonant frequency.

TABLE 13.4

Frequency	$V_{C(p-p)}$	$V_{s(p-p)}$	$I_{s(p-p)}$
10X .2ms 500 Hz	.019	3.525	
5X .2ms 1,000 Hz	.028	3.52	
10x50u 2,000 Hz	.053	3.52	
6.5X50u 3,000 Hz	.0985	3.51	
5x50u 4,000 Hz	.191	3.45	
10x20u 5,000 Hz	.483	3.06	
8.3x20 6,000 Hz	.345	3.34	
7.2x20 7,000 Hz	.191	3.48	
6.2x20 8,000 Hz	.122	3.51	
5.5x20 9,000 Hz	.090	3.53	
5x20 10,000 Hz	.072	3.53	
4982 f_p(calculated), Hz	.468	3.08	
Near { 5050	.498	3.03	
f_p { 5263	.538	2.95	

(e) Plot $V_{C(p-p)}$ on Graph 13.1 and compare with the results of part 1(h).

(f) Determine the bandwidth and cutoff frequencies from the plot of part 2(e).

BW = _____ , f_2 = _____ , f_1 = _____

(g) Is $I_{s(p-p)}$ fairly constant for the frequency range of interest?

The Transformer

OBJECT

To study the characteristics of a transformer.

EQUIPMENT REQUIRED

Resistors

$1 - 10\text{-}\Omega$, $100\text{-}\Omega$, $220\text{-}\Omega$, $1\text{-k}\Omega$

Capacitors

$1 - 100\text{-}\mu\text{F}$

Transformers

Filament — 120-V primary, 12.6-V secondary

Instruments

1 — DMM
1 — Oscilloscope
1 — Audio oscillator or function generator

EQUIPMENT ISSUED

TABLE 14.1

Item	Manufacturer and Model No.	Laboratory Serial No.
DMM		
Oscilloscope		
Audio oscillator or function generator		

TABLE 14.2

Resistors	
Nominal Value	**Measured Value**
100 Ω	99
220 Ω	220
1 kΩ	995

RÉSUMÉ OF THEORY

A *transformer* is a device that transfers energy from one circuit to another by electromagnetic induction. The energy is always transferred without a change in frequency. The winding connected to the energy source is called the *primary,* while the winding connected to the load is called the *secondary.* A step-up transformer receives electrical energy at one voltage and delivers it at a higher voltage. Conversely, a step-down transformer receives energy at one voltage and delivers it at a lower voltage. Transformers require little care and maintenance because of their simple, rugged, and durable construction. The efficiency of power transformers is also quite high.

The operation of the transformer is based on the principle that electrical energy can be transferred efficiently by mutual induction from one winding to another.

The physical construction of a transformer is shown in Fig. 14.1.

FIG. 14.1

For practical applications, the apparent power at the input to the primary circuit ($S = V_p I_p$) is equal to the apparent power rating at the secondary ($S = V_s I_s$).

That is,

$$V_p I_p = V_s I_s \qquad (14.1)$$

It can also be shown that the ratio of the primary voltage V_p to the secondary voltage V_s is equal to the ratio of the number of turns of the primary to the number of turns of the secondary:

$$\frac{V_p}{V_s} = \frac{N_p}{N_s} \qquad (14.2)$$

The currents in the primary and secondary circuits are related in the following manner:

$$\frac{I_p}{I_s} = \frac{N_s}{N_p} \qquad (14.3)$$

The impedance of the secondary circuit is electrically reflected to the primary circuit as indicated by the following expression:

$$Z_p = a^2 Z_s \qquad (14.4)$$

where $a = N_p/N_s$.

For safety reasons the applied voltage and resulting current levels will be significantly less than normally applied to a filament transformer. The result is that equations 14.1–14.4 may not be substantiated on an exact basis.

PROCEDURE

Part 1 Phase Relationship between the Transformer Primary and Secondary Voltages

(a) Set up the circuit of Fig. 14.2.

FIG. 14.2

$V_{Rms} = 4.243$

$$\frac{4.24}{.1591} = \frac{115}{12.6}$$

$.1591$

7.174

Using the alternate mode, adjust the vertical sensitivities so that both deflections are equal. What does the pattern on the screen indicate as to the phase relationship of the two voltages? Give reasons for your answer.

(b) Calculate the turns ratio from $\quad a = \dfrac{N_p}{N_s} = \dfrac{V_p}{V_s} = 12\ V_{(p\text{-}p)} = \dfrac{4.24}{152} = 8.1$

and compare to the turns ratio determined from $\quad a = \dfrac{V_p}{V_s} = \dfrac{120V}{12.6V} = 9.52$

Part 2　Voltage, Current, and Turns Ratio of the Transformer

(a) Construct the circuit of Fig. 14.3. Insert the measured resistor values. Set the input voltage to 12 V (*p-p*) using the oscilloscope.

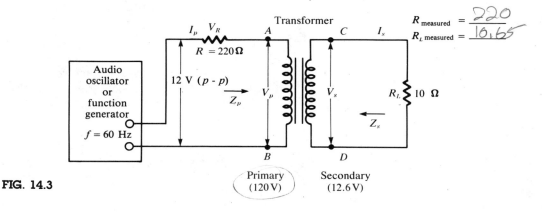

$R_{measured} = 220$

$R_{L\ measured} = 10.65$

FIG. 14.3

Primary (120 V)　Secondary (12.6 V)

Measure and record the primary and secondary rms voltage with the DMM.

$V_{p(rms)} = \underline{\quad 3.1 \quad}$, $V_{s(rms)} = \underline{\quad .355 \quad}$

Using the measured values above and Eq. (14.2), calculate the turns ratio N_p/N_s.

$a = N_p/N_s = \underline{\quad 8.79 \quad}$

(b) Determine the current I_s shown in Fig. 14.3 using Ohm's law ($I_s = V_s/R_L$). Using Eq. (14.3), calculate I_p.

$I_{s(rms)}$(measured) = _____ , $I_{p(rms)}$(calculated) = _____

(c) Determine the current I_p shown in Fig. 14.3 using Ohm's law ($I_p = V_R/R_s$).

$V_R = 2.01$

$I_{p(rms)}$ = _____

How does this measured value compare to the calculated value of (b)? Is this a step-up or step-down transformer? Explain.

(d) Set up the network of Fig. 14.4. Note the new position of terminals A through D. In essence, the transformer is hooked up in the reverse manner. Insert the measured resistor values. Set the input voltage to 4 V (*p-p*) using the oscilloscope.

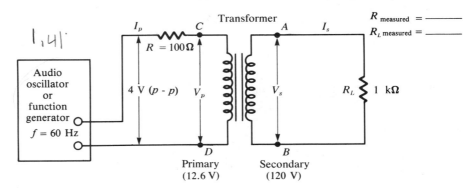

FIG. 14.4

1.41

Primary (12.6 V)   Secondary (120 V)

$R_{measured}$ = _____
$R_{L\ measured}$ = _____

Now measure the rms values of V_p, V_s with the DMM. Use the technique of the previous parts to determine I_p and I_s.

$V_{p(rms)}$ = __4.314__ , $I_{p(rms)}$ = _____
$V_{s(rms)}$ = __1.161__ , $I_{s(rms)}$ = _____

$$\frac{123}{13.47} = \frac{123}{X} = \frac{115}{12.6} \qquad 9.13$$

Using Eqs. (14.2) and (14.3), calculate the turns ratio. Is the transformer being used as a step-up or step-down transformer? Explain.

$$N_p/N_s[\text{Eq. (14.2)}] = \underline{\hspace{3cm}} , \quad N_p/N_s[\text{Eq. (14.3)}] = \underline{\hspace{3cm}}$$

Part 3   Power Dissipation

Using the measured values of I_p, V_p, I_s, and V_s, calculate the apparent power ($P = VI$) of the primary and secondary circuits of the transformer of Fig. 14.3.

$$P_{\text{primary}} = \underline{\hspace{3cm}} , \quad P_{\text{secondary}} = \underline{\hspace{3cm}}$$

What conclusion can you draw from your calculations?

Part 4   Impedance Transformation

Reconnect the transformer as shown in Fig. 14.3.

   (a) Using Eq. (14.4), calculate the reflected impedance (Z_p) for the network of Fig. 14.3. Do not include R in Z_p.

$$Z_p(\text{calculated}) = \underline{\hspace{3cm}}$$

   (b) Calculate Z_p from the data of Parts 2(a) and (c). That is, use $Z_p = V_p/I_p$.

$$Z_p = \underline{\hspace{3cm}}$$

How does this value of Z_p compare to the calculated value in (a)?

  (c) Disconnect the circuit (Fig. 14.3) from the oscillator or generator. Connect the 100-μF capacitor in series with the 10-Ω resistor. Calculate the magnitude of the reflected impedance Z_p for this circuit.

Z_p = _____

(d) Connect the circuit to the oscillator or generator. Measure I_p and V_p. ✓

I_p = _____ , V_p = _3.67_

Calculate Z_p (the magnitude) using the values of I_p and V_p above.

$V_s = .14\$$

$V_R = .11$

$V_C = .43$

Z_p = _____

How does this value of Z_p compare to the value calculated in (b)?

Pulse Waveforms

OBJECT

To study the characteristics of pulse waveforms, specifically the square wave, and the response of an R-C circuit to a square-wave input.

EQUIPMENT REQUIRED

Resistors

1 — 1-kΩ

Capacitors

1 — 0.01-μF

Instruments

1 — DMM
1 — Oscilloscope
1 — Square-wave or function generator
1 — dc Power supply

EQUIPMENT ISSUED

TABLE 15.1

Item	Manufacturer and Model No.	Laboratory Serial No.
DMM		
Oscilloscope		
Generator		
dc Power supply		

TABLE 15.2

Resistors	
Nominal Value	**Measured Value**
1 kΩ	997

RÉSUMÉ OF THEORY

Pulse Waveforms

Our analysis thus far has been limited to sinusoidal functions. In this experiment we will examine pulse waveforms and the response of an *R-C* circuit to a square-wave input.

If the switch of Fig. 15.1(a) were opened and closed at a periodic rapid rate, the voltage across the resistor *R* would appear as shown in Fig. 15.1(b). As noted in Fig. 15.1(b), if the switch is open, the output is zero volts since the current *I* is zero amperes and $V_R = IR = (0)R = 0$ V. Once the switch is closed, V_R jumps to *E* volts. The difference between the lower and upper levels of the generated pulses is called the *step amplitude*. The abrupt change in voltage level is called an electrical step. As noted in Fig. 15.1(b), electrical steps can be defined as *positive-going* or *negative-going*. The pulses of Fig. 15.1(b) are ideal pulses since the sides are shown as absolutely vertical. Unfortunately, this is not the case in practical situations. A finite time (although often very small) is required for the voltage to change levels. A few important definitions associated with pulse waveforms are provided below.

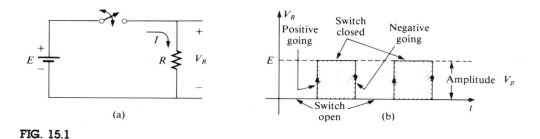

(a)   (b)

FIG. 15.1

Pulse Width (t_p):

The time between the two points on a pulse where the magnitude is one-half the maximum amplitude, as noted in Fig. 15.2.

Rise Time (t_r):

The time required for the voltage to go from $0.1V_p$ to $0.9V_p$, as noted in Fig. 15.2.

Fall Time (t_f):

The time required for the voltage to drop from $0.9V_p$ to $0.1V_p$, as noted in Fig. 15.2.

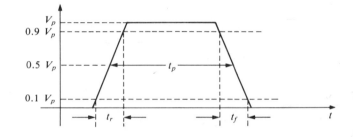

FIG. 15.2

Pulse Repetition Frequency (prf):

The rate at which the pulses of a periodic waveform repeat in a pulse train — actually the frequency of the periodic waveform. It is related to the period T by prf $= 1/T$. See Fig. 15.3.

$$t_p = 5 \text{ ms} \quad T = 10 \text{ ms}$$
$$\text{prf} = \frac{1}{T} = \frac{1}{10 \times 10^{-3}} = 100 \text{ Hz}$$

FIG. 15.3

Duty Cycle (D):

Since pulse waveforms often do not encompass 50% of the period, as shown in Fig. 15.4, the duty cycle, defined by $D = t_p/T$, is often applied to describe the pulse train.

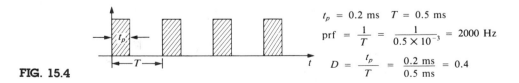

$$t_p = 0.2 \text{ ms} \quad T = 0.5 \text{ ms}$$
$$\text{prf} = \frac{1}{T} = \frac{1}{0.5 \times 10^{-3}} = 2000 \text{ Hz}$$
$$D = \frac{t_p}{T} = \frac{0.2 \text{ ms}}{0.5 \text{ ms}} = 0.4$$

FIG. 15.4

A special case of the pulse waveform is the square wave appearing in Fig. 15.5, with an average value other than zero volts. The square wave has a duty cycle of 0.5 and an average value determined by $V_{av} = (D)(V_p) +$ base-line value, where the base-line value is as indicated in Fig. 15.5.

FIG. 15.5

For the waveform of Fig. 15.5,

$$V_{av} = (D)(V_p) + V_b \tag{15.1}$$
$$= (0.5)(6) = 3 + 2$$
$$= 5 \text{ V as is confirmed by Fig. 13.5.}$$

R-C Response to a Square Wave

Consider the *R-C* circuit of Fig. 15.6 with the square-wave input. Assume that the capacitor is initially uncharged and that the signal is applied at $t = 0$ s. The mathematical expression relating the voltage $V_C(t)$ to the other parameters of the circuit is the following:

$$V_C(t) = V_f + (V_i - V_f)e^{-t/\tau} \tag{15.2}$$

where V_f is the final value of the voltage $V_C(t)$, V_i is the initial value of the voltage $V_C(t)$, and τ is the time constant of the circuit $= RC$.

FIG. 15.6

For the conditions of Fig. 15.6, $V_f = V_p$ and $V_i = 0$ V. Substituting into Eq. (15.2),

$$V_C(t) = V_p + (0 - V_p)e^{-t/\tau}$$

and

$$V_C(t) = V_p(1 - e^{-t/\tau}) \tag{15.3}$$

If we choose a frequency such that $T = 2\tau$, then t_1 of Fig. 15.6 will be defined by $t_1 = \tau$ and the voltage $V_C(t)$ will be determined by the following:

$$V_c(t) = V_p(1 - e^{-\tau/\tau}) = V_p(1 - e^{-1}) = V_p(1 - 0.367)$$
$$= 0.633V_p$$

as shown in Fig. 15.7.

FIG. 15.7

If we reduce the frequency such that $T = 10\tau$ or $t_1 = 5\tau$, then $V_C(t)$ will for all practical purposes reach its final value, as indicated in Fig. 15.8.

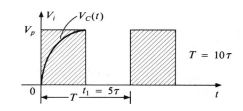

FIG. 15.8

For intermediate values of t_1 established by the applied frequency, the voltage $V_C(t)$ will reach the levels indicated in Table 15.3.

TABLE 15.3

t_1	$e^{-t/\tau}$	$(1 - e^{-t/\tau})$	$V_c(t)$
τ	0.367	0.633	$0.633V_p$
2τ	0.135	0.865	$0.865V_p$
3τ	0.049	0.951	$0.951V_p$
4τ	0.018	0.982	$0.982V_p$
5τ	0.006	0.994	$0.994V_p$

After the first positive pulse, the applied voltage abruptly drops to zero. The parameters of Eq. (15.2) then change to the following:

$$V_f = 0\,\text{V} \quad \text{and} \quad V_i = \text{value of } V_C(t) \text{ at } t = t_1$$

And substituting into Eq. (15.2) yields

$$V_c(t) = V_f + (V_i - V_f)e^{-t/\tau} = 0 + (V_i - 0)e^{-t/\tau}$$

and

$$V_c(t) = V_i e^{-t/RC} \tag{15.4}$$

For the case of $t_1 = \tau$,

$$\begin{aligned} V_c(t) &= V_i e^{-t/RC} = (0.633V_p)e^{-\tau/\tau} \\ &= (0.633V_p)e^{-1} \\ &= (0.633V_p)(0.367) \\ &= 0.232V_p \end{aligned}$$

as shown in Fig. 15.9.

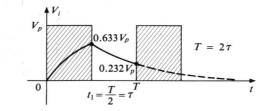

FIG. 15.9

For the second pulse, the new initial value is $0.232V_p$ with a V_f equal to V_p. In time, the waveform will settle down to that appearing in Fig. 15.10 for $t_1 = T/2 = \tau$.

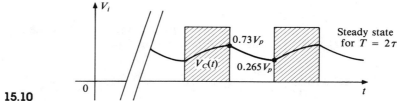

FIG. 15.10

For the case of $T = 10\tau$, the capacitor will discharge to zero and V_i will be zero volts for the second pulse, resulting in the same waveform as Fig. 15.8.

For the other possibilities of Table 15.3, the resulting waveform will be like Fig. 15.10 but with different transition values.

PROCEDURE

Part 1   Repetition Rate and Duty Cycle

(a) Set the function generator to the square-wave mode and set the frequency to 1 kHz. Connect the generator to the oscilloscope as shown in Fig. 15.11.

FIG. 15.11

Adjust the amplitude control until a 6-V (*p-p*) signal is obtained. Then adjust the sweep control until *two* full periods appear on the screen. Using the scope display, adjust the fine-control on the frequency of the function generator to insure that the output frequency is exactly 1 kHz. That is, using the horizontal sensitivity, determine how many horizontal divisions should encompass a 1-kHz signal, and make the necessary adjustments.

Determine the following for the resulting display:

pulse width = _____ .5ms _____

period = _____ 1ms _____

repetition rate = _____ 1000Hz _____

duty cycle = _____ .5 m _____

$$\frac{.5ms}{1ms}$$

Move the vertical input mode control from ac to dc and back again. Is there any change in the appearance or location of the display? Is there any average dc present? Yes There is a slight jump and Dip in the voltage

(b) Place the mode switch to ground (Gnd) and set the zero reference line in the center of the screen. Return the switch to the dc position and note that the square wave has an equal positive and negative pulse.

Determine the average value using Eq. (15.1).

$$(6p)(.5) = 3\ V_{AUG}$$

(c) Leave the generator and scope settings unchanged and construct the network of Fig. 15.12. The dc supply should be left "floating." That is, do not connect the ground terminal of the dc supply to any point in the network. Set the dc supply to 3 V and adjust the generator for a 6-V (*p-p*) signal at 1 kHz. Then set the zero reference line at the bottom of the screen and note the effect of switching the mode control from ac to dc and back. Shifts The wave up 3v

FIG. 15.12

With the mode control in the dc position, sketch the waveform and indicate the zero reference line.

(d) Using Eq. (15.1), determine the average value of the waveform of (c).

$$(.5)(6V) + 3 = 6 \text{ volts}$$

Part 2   *R-C* Response to a Square-Wave Input ($T = 100\tau$)

(a) Construct the network of Fig. 15.13.

FIG. 15.13

(b) Set the dc power supply (ungrounded) to zero volts, the square-wave generator to 1 kHz, and connect the scope to point *a*. Adjust the amplitude to 10 V (*p-p*). Using the mode switch, establish the bottom line of the screen as the zero reference level. With the mode switch on dc, adjust the power supply until the bottom of the negative half-cycle rests on the zero reference line. In addition, adjust the frequency of the generator and the horizontal sensitivity control until there are only two cycles on the screen. Draw the resulting waveform on the coordinate axis of Graph 15.1.

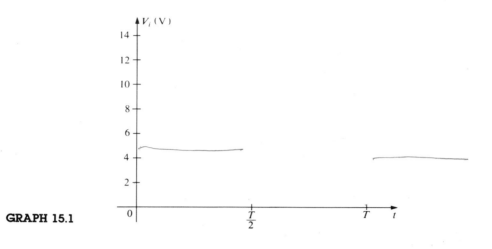

GRAPH 15.1

(c) Determine the time constant of the network. Use the measured value of R.

$10\mu s$

(d) Determine T and $T/2$ and compare to the time constant calculated in (c).

$$T = 1.ms$$
$$T/2 = 54ms$$

(e) Calculate the value of $V_C(t)$ at $t = t_1 = T/2$ using Eq. (15.3).

$$10v(1 - e^{-54}) \cong 10v$$

(f) What does the value of (e) indicate to you?

The cap is fully charged

(g) Place the scope at point b of Fig. 15.13 and draw the trace on Graph 15.2.

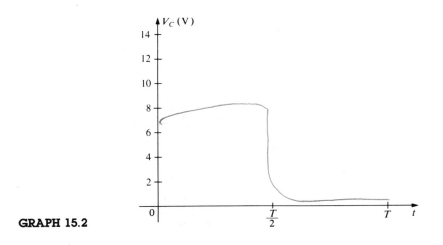

GRAPH 15.2

(h) Switch the positions of the resistor and capacitor and sketch the waveform across R on Graph 15.3. Make the adjustments necessary to insure that the total waveform is on the screen.

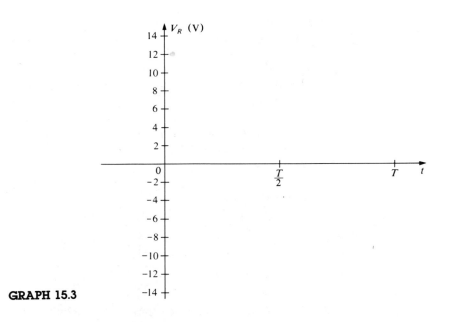

GRAPH 15.3

(i) At $t = T/4$, does $V_i = V_R + V_C$? Use the waveforms sketched above to determine V_R and V_C at $t = T/4 = 25\tau$.

Part 3 *R-C* Response for $T = 10\tau$

(a) Reconstruct the network of Fig. 15.13 and connect the scope to point *a*. This part of the experiment will examine the response for $T = 10\tau$ or $T/2 = 5\tau$. The required frequency can be determined by first finding the period T.

$T = $ _____ , $T/2 = $ _____

$f = $ _____

(b) Set the input to the same signal as in Part 2(b) but at the frequency calculated in Part 3(a). Determine the magnitude of $V_c(t)$ at $t = t_1 = T/2 = 5\tau$.

(c) Connect the scope to point *b* and sketch the waveform on the same graph used for Part 2(g). Is $V_c(t)$ the value forecasted in Part 3(b)?

(d) Switch the positions of the capacitor and resistor and sketch V_R on Graph 15.3.

(e) At $t = T/4$, does $V_i = V_R + V_c$? Use the waveforms sketched above to determine V_R and V_c at $t = T/4 = 2.5\tau$.

Part 4 *R-C* Response for $T = 2\tau$

(a) Repeat Part 3(a) for $T = 2\tau$ or $T/2 = \tau$.

$T = $ _____ , $T/2 = $ _____

$f = $ _____

(b) Set the input to the same signal as in Part 2(b) but at the frequency calculated in Part 4(a). Determine the magnitude of $V_c(t)$ at $t = t_1 = T/2 = \tau$.

(c) Connect the scope to point b and sketch the waveform on Graph 15.2. Is the steady-state waveform of $V_C(t)$ in agreement with Fig. 15.10 for $t_1 = T/2 = \tau$?

(d) Switch the positions of the capacitor and resistor and sketch V_R on Graph 15.3.

(e) At $t = T/4$, does $V_i = V_R + V_C$? Use the waveforms sketched above to determine V_R and V_C at $t = T/4 = 0.5\tau$.

(f) What is the effect of a further increase in frequency on the voltage across the capacitor or resistor? Explain why the waveform changes as it does.

Currents and Voltages in Balanced Three-Phase Systems

OBJECT

To investigate the current and voltage relations in three-phase systems with balanced Y and Δ loads.

EQUIPMENT REQUIRED

Resistors

3 — 100-Ω, 330-Ω, 18-kΩ, 33-kΩ

Instruments

1 — DMM

2 — Phase-sequence indicator

Power Supply

Three-phase, 120/208-V, 60-Hz

EQUIPMENT ISSUED

TABLE 16.1

Item	Manufacturer and Model No.	Laboratory Serial No.
DMM		
Phase-sequence indicator		

Since more than one resistor of a particular value will appear in some configurations of this experiment, the measured value of each should be determined just prior to inserting them in the network.

RÉSUMÉ OF THEORY

In a balanced three-phase system, the three-phase voltages are equal in magnitude and 120° out of phase. The same holds true for the line voltages, phase currents, and line currents. The system is balanced if the impedance of each phase is the same. For the balanced Y load, the line current is equal to the phase current, and the line voltage is the square root of 3 (or 1.73) times the phase voltage. For the balanced Δ load, the line and phase voltages are equal, but the line current is the square root of 3 (or 1.73) times the phase current.

The relationship used to convert a balanced Δ load to a balanced Y load is

$$Z_Y = \frac{Z_\Delta}{3}$$ (16.1)

Since the sum (phasor) of three equal quantities 120° out of phase is zero, if a neutral wire is connected to a balanced three-phase, Y-connected load, the current in this neutral wire will be zero. If the loads are unbalanced, the current will not be zero.

The phase sequence of voltage does not affect the magnitude of currents in a balanced system. However, if the load is unbalanced, changing the phase sequence will change the magnitude of the currents. The sequence of voltages determines the direction of rotation of a three-phase motor.

PROCEDURE

Note: Be absolutely sure that the power is turned off when making changes in the network configuration or inserting and removing instruments.

Part 1 Phase-Sequence Indicator (if available)

Connect the sequence indicator to the power source as shown in Fig. 16.1. Turn on the power. If the 1-2-3 light is brighter than the 3-2-1 light, then the phase sequence is 1-2-3 or $\mathbf{E}_{12}(\mathbf{E}_{AB})$, $\mathbf{E}_{23}(\mathbf{E}_{BC})$, and $\mathbf{E}_{31}(\mathbf{E}_{CA})$, which means that if $\mathbf{E}_{12}$ is the reference, then $\mathbf{E}_{23}$ lags $\mathbf{E}_{12}$ by 120°, and $\mathbf{E}_{31}$ lags $\mathbf{E}_{23}$ by 120°. If the 3-2-1 lamp is brighter, then the sequence is 3-2-1. Record the phase sequence of the line voltages.

Sequence = _____

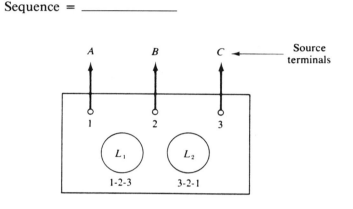

FIG. 16.1

Read the magnitude of the line voltages (E_{AB}, E_{BC}, and E_{CA}) and the phase voltages (E_{AN}, E_{BN}, and E_{CN}) using the DMM.

$E_{AB} =$ _____ , $E_{BC} =$ _____ , $E_{CA} =$ _____

$E_{AN} =$ _____ , $E_{BN} =$ _____ , $E_{CN} =$ _____

Draw the phasor diagram of the line voltages showing magnitude and phase angle if E_{AB} is the reference voltage at an angle of zero degrees.

Part 2 Resistors in a Balanced Δ

Construct the network of Fig. 16.2.

(a) Measure the three-phase voltages and line voltages.

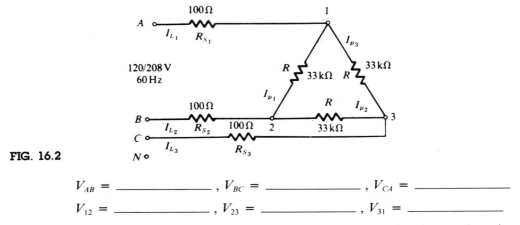

FIG. 16.2

V_{AB} = _____ , V_{BC} = _____ , V_{CA} = _____

V_{12} = _____ , V_{23} = _____ , V_{31} = _____

What is the relationship between the line voltage and the phase voltage in a balanced Δ-connected load?

(b) Measure the voltage across the sensing resistors with the DMM.

$V_{R_{S_1}}$ = _____ , $V_{R_{S_2}}$ = _____ , $V_{R_{S_3}}$ = _____

Calculate the line currents using the sensing voltages and Ohm's law.

$I_{R_{S_1}} = I_{L_1}$ = _____ , $I_{R_{S_2}} = I_{L_2}$ = _____ ,

$I_{R_{S_3}} = I_{L_3}$ = _____

(c) Using Ohm's law, and the measurements of part 2(a), calculate the phase currents.

I_{p_1} = _____ , I_{p_2} = _____ , I_{p_3} = _____

(d) For a balanced Δ-connected load, how are the line currents related (in magnitude)?

(e) For a balanced Δ-connected load, how are the phase currents related (in magnitude)?

(f) For a balanced Δ-connected load, how are the line and phase currents related (in magnitude)?

Part 3   Resistor in a Balanced Y

Construct the network of Fig. 16.3.

(a) Measure the three-phase voltages and line voltages.

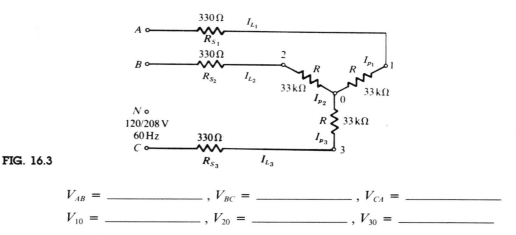

FIG. 16.3

V_{AB} = _____ , V_{BC} = _____ , V_{CA} = _____

V_{10} = _____ , V_{20} = _____ , V_{30} = _____

What is the relationship between the magnitudes of the line voltages and the phase voltages of a balanced Y-connected load?

(b) Measure the voltage across the sensing resistors with the DMM.

$V_{R_{S_1}}$ = _____ , $V_{R_{S_2}}$ = _____ , $V_{R_{S_3}}$ = _____

Calculate the line currents using the sensing voltages and Ohm's law.

$$I_{R_{S_1}} = I_{L_1} = \text{_____} , I_{R_{S_2}} = I_{L_2} = \text{_____} ,$$

$$I_{R_{S_3}} = I_{L_3} = \text{_____}$$

(c) Using Ohm's law, and the measurements of part 3a, calculate the phase currents.

$$I_{P_1} = \text{_____} , I_{P_2} = \text{_____} , I_{P_3} = \text{_____}$$

(d) For a balanced Y-connected load, how are the line currents related (in magnitude)?

(e) For a balanced Y-connected load, how are the phase currents related (in magnitude)?

(f) For a balanced Y-connected load, how are the line and phase currents related?

Part 4 Resistors in a Combination Y-Δ Configuration

(a) Construct the network of Fig. 16.4.

FIG. 16.4

(b) Read the line voltages V_{AB}, V_{BC}, and V_{CA}, and determine the line currents from the voltage across the sensing resistors.

$V_{AB} =$ ——————— , $V_{BC} =$ ——————— , $V_{CA} =$ ———————

$I_{L_1} =$ ——————— , $I_{L_2} =$ ——————— , $I_{L_3} =$ ———————

(c) Using the measured line voltage (average value) and nameplate component values, calculate a line current (magnitude) using a Y-Δ conversion, and compare it to a measured value from (b). Ignore the effects of the 100-Ω sensing resistors in your calculations.

I_L(measured) = ——————— , I_L(calculated) = ———————

(d) Repeat (c), but convert the balanced Δ to a balanced Y.

I_L(calculated) = ———————

Power Measurements in Three-Phase Systems

OBJECT

To investigate the two-wattmeter method of measuring power in three-phase systems.

EQUIPMENT REQUIRED

Resistors

3 — 250-Ω (200-W or more), wire-wound

Instruments

1 — DMM

2 — 1000-W-maximum wattmeters (HPF)

Power Supply

Three-phase, 120/208-V, 60-Hz

EQUIPMENT ISSUED

TABLE 17.1

Item	Manufacturer and Model No.	Laboratory Serial No.
DMM		
Wattmeter		
Wattmeter		

Since more than one resistor of a particular value will appear in some configurations of this experiment, the measured value of each should be determined just prior to inserting them in the network.

RÉSUMÉ OF THEORY

The total average power to a three-phase load circuit is given by Eqs. (17.1) and (17.2):

$$\boxed{P_T = 3E_pI_p \cos \theta} \tag{17.1}$$

where E_p is the phase voltage, I_p the phase current, $\cos \theta$ the power factor, and θ the phase angle between E_p and I_p.

$$\boxed{P_T = \sqrt{3}E_LI_L \cos \theta} \tag{17.2}$$

where E_L is the line voltage, I_L the line current, $\cos \theta$ the power factor, and θ the phase angle between E_p and I_p.

The power to a three-phase load circuit can be measured in two ways:

1. Three-wattmeter method
2. Two-wattmeter method

THREE-WATTMETER METHOD

The connections for the wattmeters for a Δ and a Y circuit are shown in Fig. 17.1. The total power P_T is equal to the sum of the wattmeter readings.

Note that in each case, connections must be made to the individual load in each phase, which is sometimes awkward, if not impossible. A more feasible and economical method is the two-wattmeter method.

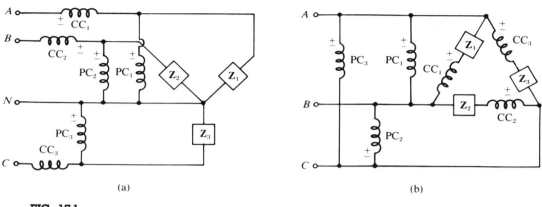

(a)   (b)

FIG. 17.1

TWO-WATTMETER METHOD

With the two-wattmeter method, it is not necessary to be concerned with the accessibility of the load. The two wattmeters are connected as shown in Fig. 17.2

FIG. 17.2

It can be shown that the two-wattmeter method of measuring total average power yields the correct value under any conditions of load.

If we plot a curve of power factor versus the ratio of the wattmeter readings P_1/P_2, where P_1 is always the smaller reading, we obtain the curve appearing in Fig. 17.3.

The graph shows that if the power factor is equal to 1 (resistive circuit), then $P_1 = P_2$ and the readings are additive. If the power factor is zero (reactive), then $P_1 = P_2$; but P_1 is negative, so $P_T = 0$. At a 0.5 power factor, one wattmeter reads zero, and the other reads the total power. It is indicative from the graph that if $F_p > 0.5$, the readings are to be added; and if $F_p < 0.5$, the readings are to be subtracted.

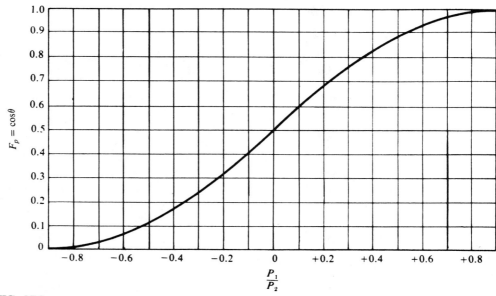

FIG. 17.3

The two-wattmeter method will read the total power in a three-wire system even if the system is unbalanced. For three-phase, four-wire systems (balanced or unbalanced), three wattmeters are needed.

It can be shown that the power read by each wattmeter is given by the following equations:

$$P_2 = E_L I_L \cos(\theta - 30°) \tag{17.3}$$

$$P_1 = E_L I_L \cos(\theta + 30°) \tag{17.4}$$

$$P_T = P_2 \pm P_1 \qquad \text{(the algebraic sum of } P_1 \text{ and } P_2, \text{ where } P_1 \text{ is the smaller reading)} \tag{17.5}$$

Another advantage of the two-wattmeter method is that it allows us to calculate the phase angle (θ) and therefore the power factor (F_p) using the expressions

$$\tan \theta = \sqrt{3}\,\frac{P_2 - P_1}{P_2 + P_1} \tag{17.6}$$

$$F_p = \cos \theta \tag{17.7}$$

PROCEDURE FOR CONNECTING THE WATTMETER USING THE TWO-WATTMETER METHOD

Connect the potential coils (PC) in the circuit last, and use the insulated alligator clip leads for their connections to line *B*. (See Fig. 17.2.) Close the breaker. If either wattmeter reads down-scale, shut down and reverse the current coil (CC) of the wattmeter that reads down-scale. Apply power again and read both wattmeters.

To find the sign of the lower-reading wattmeter, proceed as follows. If wattmeter #1 is the lower reading, unclip lead *a* of the PC coil and touch to any point on line *A*. If the wattmeter deflects up-scale, the reading is to be added (+). If the deflection is down-scale, the reading is negative and is to be subtracted (−). If wattmeter #2 is the lower-reading wattmeter, then the PC lead *a* is touched to line *C* for the test.

PROCEDURE

Part 1 Balanced Δ-Connected Resistor Loads

(a) Insert the measured values of R_1, R_2, and R_3 in Fig. 17.4.

(b) Construct the network of Fig. 17.4. Refer to the procedure for connecting the wattmeter. *Exercise caution at all times!* If in doubt, ask your instructor.

FIG. 17.4

Measure E_{AB}, E_{BC}, and E_{AC}.

$E_{AB} =$ _____ , $E_{BC} =$ _____ , $E_{AC} =$ _____

Are these the line or phase voltages of the supply? Explain.

Are these the same as the phase voltages of the load V_{AB}, V_{BC}, and V_{AC}? Explain.

(c) Calculate the magnitude of the currents I_1, I_2, and I_3.

I_1 = _____ , I_2 = _____ , I_3 = _____

Are these the line or phase currents? Explain.

Calculate the values of the currents I_A, I_B, and I_C.

I_A = _____ , I_B = _____ , I_C = _____

Are these the line or phase currents? Explain.

What is the relationship between (I_A, I_B, and I_C) and (I_1, I_2, and I_3)?

(d) Calculate the power dissipated by each resistor.

P_1 = _____ , P_2 = _____ , P_3 = _____

What is the total power?

$P_T = $ _____

Read the wattmeters.

$P_1 = $ _____ , $P_2 = $ _____

What is the total power?

$P_T = $ _____

How does this value compare to the one calculated above?

Calculate the phase angle and the power factor of the circuit.

$\theta = $ _____ , $F_p = $ _____

Compute P_1/P_2. Does this agree with the graph of Fig. 17.3? Explain.

$P_1/P_2 = $ _____

If three wattmeters were to be used, what would be the reading of each wattmeter?

$P_1 = $ _____ , $P_2 = $ _____ , $P_3 = $ _____

Part 2 Balanced Four-Wire, Y-Connected Resistor Load

(a) Construct the circuit of Fig. 17.5. Refer to the procedure for connecting the wattmeter. *Exercise caution at all times!* If in doubt, ask your instructor. Insert the measured value of each resistor.

Measure the voltages V_{AN}, V_{BN}, and V_{CN}.

$V_{AN} = $ _____ , $V_{BN} = $ _____ , $V_{CN} = $ _____

Are these the line or phase voltages? Explain.

FIG. 17.5

Measure the voltages E_{AB}, E_{BC}, and E_{AC}.

$E_{AB} = $ _____ , $E_{BC} = $ _____ , $E_{AC} = $ _____

What is the relationship between (V_{AN}, V_{BN}, and V_{CN}) and (E_{AB}, E_{BC}, and E_{AC})?

Calculate E_{AB}/V_{AN}, E_{BC}/V_{BN}, and E_{AC}/V_{CN}.

$E_{AB}/V_{AN} =$ _____ , $E_{BC}/V_{BN} =$ _____ ,

$E_{AC}/V_{CN} =$ _____

Do these ratios confirm the theory? Explain.

(b) Calculate the magnitude of the currents I_1, I_2, and I_3.

$I_1 =$ _____ , $I_2 =$ _____ , $I_3 =$ _____

Are these the line or phase currents? Explain.

How are these currents related to the magnitude of I_A, I_B, and I_C?

Calculate the magnitude of I_A, I_B, and I_C.

$I_A =$ _____ , $I_B =$ _____ , $I_C =$ _____

(c) Calculate the power delivered to the circuit.

$P_T =$ _____

Read the wattmeters.

$P_1 = $ _____ , $P_2 = $ _____

Calculate P_T from P_1 and P_2.

$P_T = $ _____

(d) Calculate the power factor F_p and the phase angle θ.

$F_p = $ _____ , $\theta = $ _____

Calculate P_1/P_2.

$P_1/P_2 = $ _____

Do the values of θ, F_p, and P_1/P_2 agree with the graph of Fig. 17.3?

(e) What effect would there be if the neutral wire were removed? Explain.

(f) Sketch the connections if three wattmeters were used to measure the total power to the network of Fig. 17.5.

Computer Analysis: Series and Parallel ac Networks

OBJECT

To apply computer analysis techniques to series and parallel sinusoidal ac networks.

EQUIPMENT REQUIRED

Computer facilities capable of applying PSPICE or language such as BASIC, PASCAL, FORTRAN, or C.

RÉSUMÉ OF THEORY

This is the first of two experiments devoted to the analysis of series and parallel networks with sinusoidal inputs. The general format of the input file is quite similar to that encountered for dc networks, with the important difference that we are now interested in the magnitude and phase of the desired signal and the applied frequency can change the general characteristics of the network. The networks to be analyzed in this experiment will have a fixed frequency, whereas those of the next experiment will examine the impact of frequency variation.

As in the dc experiments, all the networks analyzed in this experiment have been examined in earlier experiments to permit a comparison of solutions.

PROCEDURE

Part 1   Series ac Network

(a) The network of Fig. 18.1 was analyzed in Experiment ac 8. Refer to Experiment ac 8, and record the magnitude and phase angle of $\mathbf{I}$, $\mathbf{V}_R$, $\mathbf{V}_L$, and $\mathbf{V}_C$.

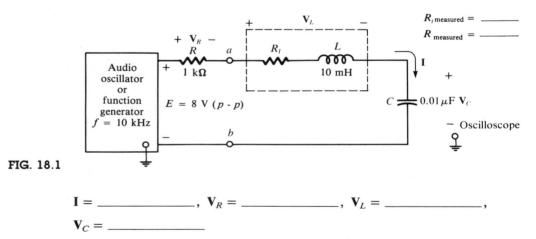

FIG. 18.1

$$\mathbf{I} = \underline{\hspace{3cm}}, \quad \mathbf{V}_R = \underline{\hspace{3cm}}, \quad \mathbf{V}_L = \underline{\hspace{3cm}},$$

$$\mathbf{V}_C = \underline{\hspace{3cm}}$$

(b) Redraw the network of Fig. 18.1, and identify the nodes as required for a PSPICE analysis.

(c) Write the input file to determine the magnitude and phase angle of **I**, $\mathbf{V}_R$, $\mathbf{V}_L$, and $\mathbf{V}_C$.

(d) Perform the PSPICE analysis and record the results.

I = _____ , $\mathbf{V}_R$ = _____ , $\mathbf{V}_L$ = _____ ,

$\mathbf{V}_C$ = _____

(e) How do the results of Part 1(d) compare with the results of Part 1(a)?

(f) Using an appropriate language, write a program to determine the magnitude and phase angle of **I**. Use a subroutine to perform the necessary polar-to-rectangular and rectangular-to-polar conversions.

(g) Run the program of Part 1(f) and record the result.

IMAG = _____ , IPHASE = _____

(h) How do the results of Part 1(g) compare with those of Part 1(a)?

Part 2   Parallel ac Network

(a) The network of Fig. 18.2 was analyzed in Experiment ac 9. Refer to the results of Experiment ac 9, and record the magnitude and phase angle of I_S, I_R, I_L, and I_C.

FIG. 18.2

$I_S =$ _____ , $I_R =$ _____ , $I_L =$ _____

$I_C =$ _____

(b) Redraw the network of Fig. 18.2, and label the network as necessary to perform a PSPICE analysis. (Ignore the effects of R_S, and use the measured value of R_l.)

      (c) Write the input file to determine the magnitude and phase angle of I_S, I_R, I_L, and I_C.

      (d) Perform the PSPICE analysis, and record the results.

$I_S =$ _____ , $I_R =$ _____ , $I_L =$ _____ ,

$I_C =$ _____

      (e) How do the results of Part 2(d) compare with those of Part 2(a)?

      (f) Using an appropriate language, write a program to determine the magnitude and phase angle of I_S. Use a subroutine for the required conversions between vector forms.

(g) Run the program of Part 2(f), and record the results.

ISMAG = _____, ISPHASE = _____

(h) How do the results of Part 2(g) compare with the values of Part 2(a)?

Computer Analysis: Frequency Response, Resonance

OBJECT

*To apply computer analysis techniques to an ac
network in which the frequency will vary
between defined limits.*

EQUIPMENT REQUIRED

Computer facilities capable of applying PSPICE
or language such as BASIC, PASCAL, FORTRAN,
or C.

RÉSUMÉ OF THEORY

This experiment will concentrate on the impact of varying the frequency applied to a network with both resistive and reactive elements. In PSPICE the PLOT routine will be used to the fullest, with the option to use .PROBE if available. For those using a language, the analysis will be limited to a tabulated listing of data points. The networks analyzed continue to be those examined in earlier experiments.

PROCEDURE

Part 1   Series R-L Network

(a) The network of Fig. 19.1 was analyzed in some detail in Experiment ac 5. Insert the measured values of R_l and R in Fig. 19.1 as obtained in Experiment ac 5.

FIG. 19.1

(b) Redraw the network of Fig. 19.1 and identify the nodes for a PSPICE analysis.

(c) Write the input file to determine the peak-to-peak values of V_L, V_R, and I for a range of frequencies from 1 kHz to 10 kHz in increments of 1 kHz. Do not be concerned about the request for peak-to-peak values rather than rms values. Simply use the peak-to-peak value as the magnitude of the applied signal, and all the other levels will also be peak-to-peak. Write the input file in a manner that will tabulate the data *and* plot the results.

(d) Perform the PSPICE analysis of Part 1(c) and compare the tabulated listing with that of Table 5.3 of Experiment ac 5. How do the results compare?

(e) Compare the PLOT response of the PSPICE analysis with the curves obtained in Graphs 5.1 and 5.2 of Experiment ac 5.

(f) Using an appropriate language, write a program to tabulate a listing of the magnitude of I(peak-to-peak) for the range of frequencies specified above.

(g) Run the program of Part 1(f) and compare the results with that appearing in Table 5.3 of Experiment ac 5.

(h) If .PROBE is available, rewrite the input file to obtain a frequency response for I, V_L, and V_R for the specified frequency range.

(i) How do the results of Part 1(h) compare to the curves of Graphs 5.1 and 5.2 of Experiment ac 5?

Part 2 Series *R-C* Network

(a) The network of Fig. 19.2 was analyzed in some detail in Experiment ac 6. Insert the measured value of R in Fig. 19.2 as obtained in Experiment ac 6.

FIG. 19.2

(b) Redraw the network of Fig. 19.2, and identify the nodes for a PSPICE analysis.

(c) Write the input file to determine the peak-to-peak values of V_C, V_R, and I for a range of frequencies from 1 kHz to 10 kHz in increments of 1 kHz. Tabulate *and* plot the results.

(d) Perform the PSPICE analysis of Part 2(c), and compare the tabulated listing with the appropriate values of Table 6.3 of Experiment ac 6. How do the results compare?

(e) Compare the PLOT response of the PSPICE analysis with the curves obtained in Graphs 6.1 and 6.2 of Experiment ac 6.

(i) How do the results of Part 2(h) compare to the curves of Graphs 6.1 and 6.2 of Experiment ac 6?

(g) Run the program of Part 2(f) and compare the results with that appearing in Table 6.3 of Experiment ac 6.

(h) If .PROBE is available, rewrite the input file to obtain a frequency response for I, V_C, and V_R for the specified frequency range.

(i) How do the results of Part 2(h) compare to the curves of Graphs 6.1 and 6.2 of Experiment ac 6?

Part 3   Series Resonance

(a) The series resonant circuit of Fig. 19.3 was analyzed in some detail in Experiment ac 12. Insert the measured values of R_1 and R in Fig. 19.3 as obtained in Experiment ac 12.

(b) Redraw the network and identify the nodes as required for a PSPICE analysis.

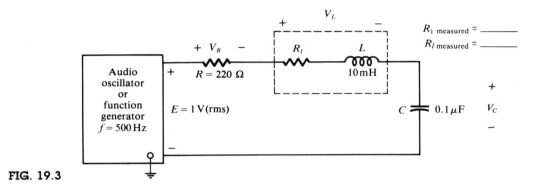

FIG. 19.3

(c) Write the input file to tabulate and plot the magnitude of I(rms) versus frequency for a frequency range of 500 Hz to 10 kHz in increments of 500 Hz.

(d) Perform the PSPICE analysis, and compare the results obtained with the data of Table 12.3 of Experiment ac 12 (for the frequencies that appear in both listings). Also, compare the resulting plot with the curve of Graph 12.1 of Experiment ac 12.

(e) Using an appropriate language, write a program to tabulate a listing of the rms value of I for the indicated range of frequencies with the same increment.

(f) Run the program of Part 3(e), and compare the results with corresponding frequencies of Table 12.3 of Experiment ac 12.

(g) If .PROBE is available, rewrite the input file to obtain a frequency response for I for the specified frequency range.

(h) How do the results of Part 3(g) compare with the curve of Graph 12.1 of Experiment ac 12?

Parts List for the Experiments

dc (REQUIRED)

Resistors

1/2-W carbon
    1—10 Ω, 47 Ω, 91 Ω, 470 Ω, 22 kΩ, 1 MΩ
    2—220 Ω, 1.2 kΩ, 2.2 kΩ, 10 kΩ, 100 kΩ
    3—100 Ω, 330 Ω, 1 kΩ, 3.3 kΩ
    1—unknown between 47 Ω and 220 Ω
1-W carbon
    1—1 MΩ
2-W carbon
    1—1 MΩ
Variable
    1—(0–1 kΩ) potentiometer, linear, carbon
    1—(0–10 kΩ) potentiometer, linear, carbon
    1—(0–250 kΩ) potentiometer, linear, carbon

Capacitors

1—1 μF
2—100 μF (electrolytic)

Inductors

1—10 mH

Miscellaneous

1—single-pole, single-throw switch
1—1 mA, 1000 Ω meter movement
1—NE-2 neon glow lamp

Instruments

1—VOM
1—DMM
1—dc Power supply

dc (OPTIONAL)

Second dc power supply
Commercial Wheatstone bridge

ac (REQUIRED)

Resistors

1/2-W carbon
    $1 - 10\ \Omega$, $22\ \Omega$, $33\ \Omega$, $47\ \Omega$, $220\ \Omega$, $470\ \Omega$, $1\ k\Omega$, $1.2\ k\Omega$, $2.2\ k\Omega$, $3.3\ k\Omega$
    $3 - 10\ \Omega$, $100\ \Omega$, $330\ \Omega$, $10\ k\Omega$, $18\ k\Omega$, $33\ k\Omega$
200-W-wirewound
    $3 - 250\ \Omega$ (Experiment 17 only)
Variable
    $1 - (0$–$1\ k\Omega)$ potentiometer
    $1 - (0$–$10\ k\Omega)$ potentiometer

Capacitors

$1 - 0.0047\ \mu F$, $0.01\ \mu F$, $0.1\ \mu F$, $0.5\ \mu F$, $1\ \mu F$, $15\ \mu F$
$2 - 0.02\ \mu F$

Inductors

$1 - 1\ mH$
$2 - 10\ mH$

Miscellaneous

1—Transformer, 120 V primary, 12.6 V secondary
2—D batteries and holders

Instruments

1—DMM (or VOM)
1—Audio oscillator or function generator
1—General purpose oscilloscope
1—dc Power supply
1—Function generator (Experiment 15 only)
2—1000 W HPF watt meters (Experiment 17 only)

Power Sources

Three-phase 120/208-V supply (For Experiments 16 and 17 only)

ac (OPTIONAL)

1—Phase sequence indicator

dc AND ac (TOTAL REQUIREMENTS)

Resistors

1/2-W carbon
    1—10 Ω, 22 Ω, 33 Ω, 47 Ω, 91 Ω, 470 Ω
    2—220 Ω, 1.2 kΩ, 2.2 kΩ, 100 kΩ
    3—100 Ω, 330 Ω, 1 kΩ, 3.3 kΩ, 10 kΩ, 18 kΩ, 33 kΩ
    1—unknown between 47 Ω and 220 Ω
1-W carbon
    1—1 MΩ
2-W carbon
    1—1 MΩ
200-W-wirewound
    3—250Ω (Experiment 15 only)
Variable
    1—(0–1 kΩ) potentiometer, linear, carbon
    1—(0–10 kΩ) potentiometer, linear, carbon
    1—(0–250 kΩ) potentiometer, linear, carbon

Capacitors

1—0.0047 μF, 0.01 μF, 0.1 μF, 0.5 μF, 1 μF, 15 μF
2—0.02 μF, 100 μF (electrolytic)

Inductors

1—1 mH
2—10 mH

Miscellaneous

1—Single-pole, single-throw switch
1—1 mA, 1000Ω meter movement
1—Transformer, 120 V primary, 12.6 V secondary
2—D batteries and holders

Instruments

1—VOM
1—DMM
1—dc Power supply
1—Audio oscillator or function generator
1—General purpose oscilloscope
1—Function generator (Experiment 15 ac only)
2—1000 W HPF wattmeters (Experiment 17 ac only)

Power Sources

Three-phase 120/208-V supply (Experiments 16 ac and 17 ac only)

dc AND ac (OPTIONAL)

Second dc power supply
Commercial Wheatstone bridge
Phase sequence indicator

Appendix I

Resistor Color Coding

1 First significant digit
2 Second significant digit
3 Multiplying value
4 Tolerance (percent)

FIG. I.1

TABLE I.1   RETMA color code for resistors

Color	1 First Significant Digit	2 Second Significant Digit	3 Multiplier	4 Percent Tolerance
Silver			0.01	10
Gold			0.1	5
Black		0	1.0	
Brown	1	1	10	1
Red	2	2	100	2
Orange	3	3	1,000	3
Yellow	4	4	10,000	4
Green	5	5	100,000	
Blue	6	6	1,000,000	
Purple	7	7	10,000,000	
Gray	8	8	100,000,000	
White	9	9	1,000,000,000	
No color				20

EXAMPLE I.1   Application of resistor color code

Nominal Value	Percent Tolerance	Band				Tolerance	Range
		1	2	3	4		
22,000 Ω	±20	Red	Red	Orange	No Band	±4400 Ω	17,600–26,400 Ω
100 Ω	±5	Brown	Black	Brown	Gold	±5 Ω	95–105 Ω
10 Ω	±10	Brown	Black	Black	Silver	±1 Ω	9–11 Ω
1.2 kΩ	±5	Brown	Red	Red	Gold	±60 Ω	1140–1260 Ω
6.8 MΩ	±20	Blue	Gray	Green	No Band	±1.36 MΩ	5.44–8.16 MΩ

Capacitor Color Coding

MOLDED MICA CAPACITORS

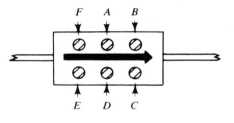

A First significant digit
B Second significant digit
C Multiplying value
D Tolerance (percent)
E Characteristic
F For Mil-C-5 A color code, dot is always black. For EIA color code, dot is always white.

FIG. II.1

TABLE II.1   Color code for molded mica capacitors

Color	A First Significant Digit	B Second Significant Digit	C Multiplier	D Percent Tolerance
Black	0	0	1	20
Brown	1	1	10	1
Red	2	2	100	2
Orange	3	3	1,000	3
Yellow	4	4	10,000	4
Green	5	5	100,000	5
Blue	6	6	1,000,000	6
Violet	7	7	10,000,000	7
Gray	8	8	100,000,000	8
White	9	9	1,000,000,000	9
Gold			0.1	5
Silver			0.01	10

MOLDED CERAMIC CAPACITORS

A　First significant digit
B　Second significant digit
C　Multiplying value
D　Tolerance
G　Temperature coefficient

FIG. II.2

TABLE II.2　　Color code for molded ceramic capacitors

Color	A First Significant Digit	B Second Significant Digit	C Multiplier	D Tolerance Value Greater Than 10 pF (%)	Value 10 pF or Less (pF)	G Temperature Coefficient (parts per million per °C)
Black	0	0	1	20	2.0	0
Brown	1	1	10	1		−30
Red	1	1	100	2		−80
Orange	3	3	1,000			−150
Yellow	4	4	10,000			−220
Green	5	5	100,000	5	0.5	−330
Blue	6	6	1,000,000			−470
Violet	7	7	10,000,000			−750
Gray	8	8	0.01		0.25	+30
White	9	9	0.1	10	1.0	+550

Appendix III

Resistance Measurements (VOM)

Never make resistance measurements on a live circuit!

Select the two leads used for resistance measurements. (Consult the instruction manual.) Set the function selector switch to the ohms position. Select the proper range with the range switch. The ranges are usually marked as multiples of R. For example,

$$R \times 1 \qquad R \times 10 \qquad R \times 100 \qquad R \times 1\,k\Omega$$

The value of the resistor can be found by multiplying the reading by the range setting. For example, a reading of 11 on the $R \times 1\,k\Omega$ range is $11 \times 1\,k\Omega = 11\,k\Omega$, or $11{,}000\,\Omega$. Note that the ohms scale (usually the topmost scale) reads from right to left, opposite to the other scales on the meter face.

Before you attach the unknown resistance to the test leads, short the leads together and observe the reading. If the meter does not read zero ohms, adjust the ZERO ADJUST control until it does. Disconnect the leads. The meter should now read infinite resistance (the extreme left-hand index of the scale). If the meter does not read infinite ohms, an adjustment will be required, using the OHMS ADJUST control. On the VOM, the control is a mechanical adjustment which requires the use of a screwdriver. Do not attempt this adjustment on the VOM without first consulting your instructor. The zero and infinite ohm adjustments must be checked each time the range of the ohmmeter is changed. Otherwise the readings will be incorrect.

Appendix IV

Proper Use of the VOM Multimeter

1. Always start with the highest range of the instrument and switch down to the proper range successively.
2. Use the range in which the deflection falls in the upper half of the meter scale.
3. Whenever possible, choose a voltmeter range such that the total voltmeter resistance (ohms/volt × FSD) is at least 100 times the resistance of the circuit under test. This rule will prevent erroneous readings due to the loading of the circuit under test.
4. Use an ohmmeter range so that the deflection falls in the uncrowded portion of the scale.
5. Exercise extreme caution when measuring voltages and currents in high-voltage circuits. A safe procedure is to shut down the power, connect the meter, turn the power on, read the meter, and again shut down to disconnect.
6. Try to ascertain the polarity of dc voltages before making the measurement.
7. Never measure resistances in a live circuit. Always shut down before making resistance measurements.
8. Whenever measuring the resistance of a resistor in a circuit, note whether there are any other resistive elements that could cause an error in the reading. It may be necessary to disconnect one side of the resistor before measuring.
9. Check the zero and ohms adjustments each time the range is changed. For proper procedure, see Appendix III.
10. When making measurements, grip the test prods by the handles as close to the lead end as possible. *Do not allow the fingers to touch the prod tips while measuring.*
11. When the instrument is not being used, do not leave it in the ohmmeter function. If the instrument has an OFF position, use it; otherwise, switch to its highest dc voltage range.
12. *Keep the instruments away from the edge of the workbench, and away from heat and dangerous fumes.*

Appendix V

Scientific Notation and Trigonometric Identities

Scientific Notation

The positive exponent of 10 is equal to one less than the number of places to the left of the decimal point:	The negative exponent of 10 is equal to the number of places to the right of the decimal point:
$10^0 = 1.0$	$10^0 = 1.0$
$10^1 = 10.0$	$10^{-1} = 0.1$
$10^2 = 100.0$	$10^{-2} = 0.01$
$10^3 = 1000.0$	$10^{-3} = 0.001$
$10^4 = 10000.0$	$10^{-4} = 0.0001$
$10^5 = 100000.0$	$10^{-5} = 0.00001$
etc.	etc.

Dimensional Prefixes

Prefix	Multiplier
tera	10^{12}
giga	10^{9}
mega	10^{6}
kilo	10^{3}
hecto	10^{2}
deka	10^{1}
deci	10^{-1}
centi	10^{-2}
milli	10^{-3}
micro	10^{-6}
nano	10^{-9}
pico	10^{-12}

Trigonometric Identities

$$\sin(-\theta) = -\sin\theta$$
$$\cos(-\theta) = \cos\theta$$
$$\tan(-\theta) = -\tan\theta$$
$$\sin(\theta + 90°) = \cos\theta$$
$$\cos(\theta - 90°) = \sin\theta$$
$$\sin^2\theta + \cos^2\theta = 1$$
$$\cos(A + B) = \cos A \cos B - \sin A \sin B$$
$$\cos(A - B) = \cos A \cos B + \sin A \sin B$$
$$\sin(A + B) = \sin A \cos B + \cos A \sin B$$
$$\sin(A - B) = \sin A \cos B - \cos A \sin B$$